CONSUMING GEOGRAPHIES

•

Food occupies a seemingly mundane position in all our lives, yet the ways we think about shopping, cooking and eating are actually intensely reflexive. The daily pick and mix of our eating habits is one way in which we experience spatial scale. From the relationship of our food intake to our body-shape to the impact of our tastes upon global food-production regimes, we all read food consumption as a practice which impacts on our sense of place.

The 'cultural turn' has brought identity politics and issues of consumption to the fore in geography. This book aims to ground theoretical debates about identity politics and issues of consumption through an exploration of one of the most universal and mundane features of everyday life. Using the example of food to demonstrate the importance of space and place in identity formation, this book contributes a geographical perspective to cultural studies work in this field.

The geography of food has thus far concentrated on production. At the same time, the emerging geographies of consumption have only mapped certain consumption sites (notably shopping malls) or particular consumption practices (the consumption of the visual landscape, for example). There is a need to expand on this, and think about the geographies of food consumption. This book considers geographies of food consumption through spatial scales, moving from the body up to the global. It focuses on the social and cultural meanings of food consumption, including material on shopping, cooking and eating; food rituals and etiquettes; food technology and the food media (TV and radio cookery programmes, food magazines, advertising); as well as selected production practices (home growing, for instance). Drawing on literatures from anthropological, sociological and cultural readings of food consumption, as well as empirical material from research into domestic food consumption practices, Bell and Valentine articulate the relationship between food and geography.

By exploring geographies of food consumption, we can begin to unpack the role food plays in constituting place identities. We all think place (and) identity through food: we are *where* we eat.

David Bell is a Lecturer in Cultural Studies at Staffordshire University. **Gill Valentine** is a Lecturer in Geography at the University of Sheffield.

CONSUMING GEOGRAPHIES

We are where we eat

•

David Bell and Gill Valentine

London and New York

First published 1997
by Routledge
11 New Fetter Lane, London EC4P 4EE

Simultaneously published in the USA and Canada
by Routledge
29 West 35th Street, New York, NY 10001

Typeset in Garamond by
Solidus (Bristol) Limited

Printed and bound in Great Britain by Butler & Tanner Ltd, Frome and London

British Library Cataloguing in Publication Data
A catalogue record for this book is available from the British Library

Library of Congress Cataloguing in Publication Data
Bell, David
Consuming geographies: we are where we eat/David Bell and Gill Valentine.
p.   cm.
Includes bibliographical references and index.
1. Food habits–Cross-cultural studies.   2. Food consumption–Cross-cultural studies.
I. Valentine, Gill.   II.   Title.
GT2850.B45   1997
394.1'2–dc20   96-44153

ISBN 0-415-13767-5
0-415-13768-3 (pbk)

CONTENTS

•

FIGURES AND PLATES

•

FIGURES

PLATES

BOXES

•

ACKNOWLEDGEMENTS

•

We would like to thank Milton Montenego/Imaginario for the wonderful cover photograph; David Maddison from the Photographic Unit at the Department of Geography, University of Sheffield for taking Plate 2.1; Corbis-Bettmann for supplying Plates 2.2, 2.3, 3.1, 3.2; CNT Marketing for Plate 4.2; Hannah Taylor from the Department of Geography at the University of Sheffield for taking Plate 5.2; BBJ for Plate 5.1; Burkitt Edwards Martin Ltd for Plate 6.2; Steve Bell for Plate 7.1 and *Sainsury's Magazine* for Plates 7.2 and 8.1. Graham Allsopp from Cartographic Unit at the Department of Geography, University of Sheffield for redrawing Figure 6.1; Peter Harrison for his copy-editing; Tamsin Meddings for her proofreading; and Anne Owen, the desk editor, for all her efforts in perfecting the final product.

We are grateful to the following for permission to reproduce their work: David Austin of the *Guardian* newspaper (Figure 2.1); Andrew Williamson of the Richard Stone agency (Plate 5.2); and Nik Huffnan from *Globehead!*, PO Box 10376, State College, PA 16805, USA, for the recipes of the centurions. Plate 6.3 appears by kind permission of Olivio. The quotation from the screenplay of *Pulp Fiction* by Quentin Tarantino, which opens the Introduction, appears by kind permission of Faber and Faber Ltd. The extracts in Chapter 2 from Jenefer Shute's novel *Life Size*, copyright © 1992 by Jenefer Shute, are reprinted by kind permission of Houghton Mifflin Company. All rights reserved.

We also wish to thank Shaun Jefferies of the *Guardian* newspaper for generously helping us to trace the holders of the copyright for some of the advertising images featured in this book.

Every attempt has been made to obtain permission to reproduce copyright material. If any proper acknowledgement has not been made, we would invite copyright holders to inform us of the oversight.

We are grateful to the Leverhulme Trust for funding the research project (award number F118AA) from which the boxed material that features in each chapter is drawn. We also wish to thank Beth Longstaff for all her hard work, efficiency and enthusiasm while employed as the Research Assistant on this project.

The recipes that open each chapter in this book are taken from the wonderful series 'recipes of the Centurions' which runs in the American publication *Globehead! A Journal of Extreme Geography*. Responding with wit and flair to the publication of a list of geography's 'centurions' (the most cited players according to citation analysis by Andrew Bodman (1992)), *Globehead!* wrote to the champions of citation and asked for their favourite recipes. We are most grateful to *Globehead!* for permission to reprint these; we are sure you'll agree they shed new light on the food–geography equation. (In some cases we have abridged the recipes; full versions, often with accompanying photographs of the chef-authors, are published in *Globehead!*) We include the centurions' ranking in Bodman's table, but would urge readers to re-rank them on the basis of their culinary creations.

Thanks also go to Tim Edensor and Cultural Studies students at Staffordshire University (especially Dani Stephenson), members of staff in the Division of Geography, Staffordshire University (especially Paul Draycott and John Ambrose) and (again) Beth Longstaff at the Department of Geography, University of Sheffield for many helpful and enjoyable discussions about food's geographies.

We owe a special debt of gratitude to Tristan Palmer for his support, and especially for commissioning this book, and to Sarah Lloyd for her enthusiastic commitment to seeing the project through.

Finally, we would like to acknowledge the support of the people that we have shared special meals with in the course of writing this book: Ruth, Daisy, Julianne, Deborah, Sarah, Liz, Jon and Julia.

REGINALD GOLLEDGE (31)
Globehead Striped Bass

•

This recipe was first conceived and enjoyed in Santa Barbara in 1990. It was prepared for Professor Gunnar Olsson after he delivered the sixth Golledge Lecture, entitled 'Forms of Thought'. The dish is simple to prepare and delicious to eat; the recipe is mine but is a variation on one developed by my spouse for preparing salmon steaks.

INGREDIENTS (SERVES FOUR)

2 lb of freshly caught striped bass fillets
1 cup of good quality chardonnay
1 stick of butter
4 full heads of garlic
Capers
Fresh ground black pepper
Freshly picked sun-ripened Goleta lemon

PROCEDURE

1   Organise a trip to Lake Havasu, Lake Mead, Lake Nacimiento, or some other Western lake that is stocked with striped bass.
*   Catch two bass, 3–5 lb each.
*   Fillet, removing all bones and skin.
*   Hurry home.

Note: Western striped bass are recommended because they are not an endangered species (like East Coast striped bass of Marxist geographers). You most probably will find the fishing trip to the West positivistically exciting!

2   Cut fillets into half-pound sections.
*   Clarify the stick of butter and brush the upper side of each fillet thoroughly with clarified butter.
*   Slowly drip a quarter of a cup of Chardonnay over each fillet.
*   Peel garlic, separate into individual cloves, slice each clove on the bias and sprinkle liberally across the top of each fillet.

Warning: Do not crush garlic. The thin roasted slices add considerably to the taste and texture of the dish.

3   Grind black pepper across the top of each fillet to taste. Green or red peppercorns can be substituted or all three can be combined into a single mixture. Pepper generously.
4   Sprinkle 2–3 teaspoons of capers over the top of each fillet.
5   Artistically arrange thin slices of lemon across the top of each fillet.
6   Prepare for broiling. Brush broiling pan lightly with the remains of the clarified butter before placing fillets in the pan under the broiler. Cook to taste. Depending on the thickness of fillets, cooking time will vary. There is plenty of room for controlled experimentation in the true spirit of scientific method.
7   Serve with garnish of sliced lemons and fresh parsley sprigs.
8   Beverage: Do not under any circumstances serve Canadian, Midwestern, or Eastern US wines with this food. While these wines may at times be used in the preparation of food, I understand it is not recommended that you drink them. I would suggest a modest Californian Chardonnay of relatively recent vintage, such as Acadia or a Château Montelana.
9   For those who are not interested in pursuing the science of cooking I suggest you adopt a postmodernist approach and imagine that you have caught, prepared and eaten the bass.

1

INTRODUCTION

•

Two men, Jules Winnfield and Vincent Vega, are driving through Hollywood in a 1974 white Chevy Nova. As they drive, Vincent tells Jules about his trip to Europe. After discussing Amsterdam's hash bars, the conversation moves on . . .

Vincent: But you know what the funniest thing about Europe is?

Jules: What?

Vincent: It's the little differences. I mean, they got the same shit over there that we got here, but it's just, just, there it's a little different.

Jules: Example.

Vincent: Well, you can walk into a movie theater and buy a beer. And I don't mean just, like, in no paper cup. I'm talking about a glass of beer. And in Paris, you can buy a beer at McDonald's. And, you know what they call a Quarter-Pounder with Cheese in Paris?

Jules: They don't call it a Quarter-Pounder with Cheese?

Vincent: No, man, they got the metric system there, they wouldn't know what the fuck a Quarter-Pounder is.

Jules: What'd they call it?

Vincent: They call it a Royale with Cheese.

Jules: Royale with Cheese.

Vincent: Yeah, that's right.

Jules: What'd they call a Big Mac?

Vincent: Well, Big Mac's a Big Mac, but they call it Le Big Mac.

Jules: Le Big Mac. What do they call a Whopper?

Vincent: I dunno, I didn't go into a Burger King. But you know what they put on French fries in Holland instead of ketchup?

Jules: What?

Vincent: Mayonnaise.

Jules: Goddamn!

Vincent: I seen 'em do it, man. They fuckin' drown 'em in that shit.

Jules: Yuck.

This dialogue comes from the screenplay of Quentin Tarantino's movie *Pulp Fiction* (1994: 14–16). Jules and Vincent are two hitmen on their way to a job, idly chatting as they go. But this part of their conversation contains some very resonant discussions of food, place and identity – topics which are at the heart of this book. An American in Paris, eating at McDonald's, is someone (whether self-consciously or not) who is connecting with 'home' while also experiencing the 'little differences' at play within an increasingly globalised culture of cuisine. And being able to get a beer in a French cinema or fast-food eatery immediately signals very different legal and cultural attitudes towards alcohol, while Jules' and Vincent's shared revulsion at the idea of mayonnaise on French fries (Vincent calls it 'that shit', Jules' only comment is 'Yuck') shows how food practices code and demarcate cultural boundaries, even in the most mundane of ways. So, in a scene which lasts no more than a couple of minutes, Tarantino captures a whole set of ideas embodied in Uma Narayan's discussion, in a more scholarly context, of what has been characterised as 'thinking about (or through, or with) food':

> Thinking about food has much to reveal about how we understand our personal and collective identities. Seemingly simple acts of eating are flavoured with complicated and sometimes contradictory cultural meanings. Thinking about food can help reveal the rich and messy textures of our attempts at self-understanding, as well as our interesting and problematic understandings of our relationship to social Others.
>
> Narayan 1995: 64

THINKING THROUGH FOOD

For most inhabitants of (post)modern Western societies, food has long ceased to be merely about sustenance and nutrition. It is packed with social, cultural and symbolic meanings. Every mouthful, every meal, can tell us something about our selves, and about our place in the world. As Arjun Appadurai (1981: 494) says, food is both 'a highly condensed social fact' and a 'marvelously plastic kind of collective representation' with the 'capacity to mobilize strong emotions'. And in a world in which self-identity and place-identity are woven through webs of consumption, what we eat (and where, and why) signals, as the aphorism says, who we are (for critical discussion of theories of the relationship between self-identity and consumption, see Warde 1994).

By focusing on a set of commonplace consumption practices – food shopping, cooking, eating and drinking – we can begin to think about a whole set of contemporary social and cultural issues, from health to nationalism, from ethics to aesthetics, from local politics to the role of transnational corporations in global regimes of accumulation. It might seem that looking at an apple, or a microwave, or a cookbook, or a supermarket is no more than epistemological or methodological slackness, but in fact, if we look at a powerful and compelling parallel work, which unpacks a similarly mundane consumption practice – smoking – then we can begin to see how this kind of pop-cultural criticism opens up rich theoretical insights; and if we then look at the ways in which food has been used in academic inquiry by historians, sociologists, philosophers and others, we can start to appreciate just how telling thinking about (or through) food can be.

Richard Klein's *Cigarettes Are Sublime* (1993) uses techniques of literary criticism and (popular) cultural analysis to reflect, with some verve and panache, on smoking. Where many texts on the subject offer dry polemic on the evils of the global tobacco industry, or on health issues, Klein takes his readers on a smoke-spiral journey through the history of cultural representations of cigarette smoking – or what he calls 'cigaretticism'. Early on in the book, Klein sets out his stall by declaring cigarettes 'America's gift to the world', praising their 'remarkable gift to modernity' (14). Later, tracking the history of tobacco, he suggests that Columbus brought tobacco from the New World as an antidote to the anxieties which his discoveries of 'a great unknown world' provoked in the 'Eurocentred consciousness of Western culture' (27), while in a survey of literary allusions to smoking he notes that poets often use cigarettes as 'powerful instruments for appropriating the world symbolically' (52). And given that something like a third of all adults have, for the past century, smoked cigarettes daily, their cultural significance is indeed remarkable.

As with cigarettes, so it is with food, with cooking, with eating and drinking. Even more universal and commonplace, seemingly even more mundane, food in fact occupies an unrivalled centrality in all our lives; although it might not dominate our consciousness, it nevertheless serves to structure our lives, from the daily rhythm of meals to the rites of consuming passage (our first sip of alcohol, the first time we cook a meal, right up to our 'last supper'). Arjun Appadurai (1993) has theorised these links, attempting to 'resituate consumption in time' (11), beginning with consumption's role in punctuating or periodising all our lives through repetitive 'techniques of the body'. Consumption – and here an explicit focus on eating is called into play – 'calls for habituation, even in the more upscale environments where food has become largely dominated by ideas of bodily beauty and comportment, rather than by ideas of energy and sufficiency' (12). These daily rhythms impact on other, longer-duration periodisations of consumption:

> In any socially regulated set of consumption practices, those that center around the body, and especially around the feeding of the body, take on the function of structuring temporal rhythm, of setting the minimum temporal measure (by analogy to musical activity) on which much more complex, and 'chaotic' patterns can be built. Pushing the analogy a step further, the small habits of consumption, typically daily food habits, can perform a percussive role in organizing large-scale consumption patterns, which may be contrived of much more complex orders of repetition and improvisation.
>
> Appadurai 1993: 13

Seasonality and rites of passage are thus also marked by consumption practices – or, rather, are built from them, as, at all scales, 'consumption creates time' (15) rather than merely responding to it; hence 'natural' periodisations – including 'seasons' and punctuations in the life course – can better be seen as *naturalised* 'consumption seasonalities' (16). Later in the essay, Appadurai builds on Thompson's (1967) classic account of the rise of time discipline which occurred with industrialisation. Comments on the resultant commodification of time (and especially the demarcation of 'work time' and 'leisure time') build to an account of the increasing of time discipline in consumption (what Gofton (1990) calls 'time famines'), such that consumption itself becomes work, with everyone

> laboring daily to practice the disciplines of purchase, in a landscape whose temporal structures have become radically polyrhythmic. Learning these multiple rhythms (of bodies, products, fashions, interest rates, gifts, and styles) and how to interdigitate them is not just work, it is the hardest sort of work, the work of the imagination.
>
> Appadurai 1993: 31

Contemporary consumption, Appadurai concludes, is governed by ephemerality, scopophilia and body manipulation linked in a systematic and generalised way into 'a set of practices that involve a radically new relationship between wanting, remembering, being and buying' (33). Drawing on Emily Martin's (1992) combining of theories of corporeality and flexibility, he suggests that bodies, consumption, fashion and time come together in the following logic: 'ephemerality becomes the civilizing counterpart of flexible accumulation and the work of the imagination is to link the ephemerality of goods with the pleasures of the senses' (33).

In a broadly similar way, Leslie Gofton (1990) mobilises Thompson's work on time discipline and the typology pre-industrial/industrial/post-industrial in a discussion of food's changing role in our experience of time. For Gofton, food technologies have radically altered our relationship to eating:

> Food itself is considered less 'significant', and carries less symbolic weight than in the past. It doesn't signify the season, or the time of day, or the day of the week in quite the way it did, nor does it mark out the roles and relations between adults and children within formal meals.
>
> Gofton 1990: 92

The household is now flexible and self-reliant (at part under current state-defined notions such as 'active citizenship'), time-budgeting has become domesticated, and so on. We have, he concludes, '[i]nformal, open-ended, anomic food choice' (87), with our total reliance on outside production giving us ever greater anxiety about food safety and acceptability.

FOOD AS POPULAR CULTURE

One way of thinking about food is as part of contemporary popular culture – especially given the fact that commentators increasingly see consumption as the dominant contemporary cultural force. Steve Redhead (1995) argues that a global popular culture industry has developed, incorporating many previously disparate areas of leisure and pleasure – together with the cultural commentaries that accompany them. Food has become very much a part of this industry, woven into the construction of 'lifestyles' (Tomlinson 1990) and used as a marker of social position (Bourdieu 1984). The food media have been instrumental in this, and recent years have seen a proliferation of food professionals, mediatisers and celebrities. Professional and amateur chefs are household names (British TV shows like *Ready Steady Cook*, *Masterchef* and *Food and Drink* bring *haute cuisine* into our living-rooms as well as our kitchens), their restaurants given the status of temples of consumption in countless guides and features; food writers, critics and broadcasters meanwhile show us not only how to cook, but tell us what, when, where, how – and even why – to eat and drink. We might even go so far as to argue that the food media make stars of the foodstuffs themselves, investing in them such cultural capital that they take on meanings far away from mere ingredients, recipes and dishes. Food magazines contain luscious centre-folds (referred to quite accurately by Barry Smart (1994) as 'gastro-porn' – seductive but unobtainable and artificially, stereotypically perfect), adverts in print and on TV offer us all manner of temptation, tying food in with all the old favourites: with health, with status, with sex, with existential happiness, or *jouissance*. Everyone is catered for, in cookbooks from *Zen Cookery* to *White Trash Recipes*; we can learn to 'cook for pleasure' (for friends and family, or just for ourselves – remember Delia Smith's *One Is Fun!*) or to 'cook in a hurry' (*Microwave Meals in Minutes*); we can create a macrobiotic diet, learn about food combining, or find 'a thousand ways with mince'.

One recurring theme in the food media – which we will be paying close attention to later – is the sampling of other cultures through their food. A British Sunday paper thus offers us insights into the 'global kitchen', and we can buy countless books like *Around the World in 80 Dishes*, offering 'authentic' recipes, menus and techniques, from TexMex to Thai. Vendors and advertisers increasingly appeal to what the food industry calls 'geographical product descriptors' (Hodgson and Bruhn 1993) in an effort to cash in on this phenomenon, while the potential offered by the Internet for 'kitchen table tourism' has been readily seized – there is an incredible listing of food interest groups on the information superhighway, with every specialism from insect recipes to championship chillies and 'Mimi's Cyber Kitchen'; the University of Guadalajara offers access via soundcard to an audio version of an encyclopedic Spanish cookbook which it recently translated into English. Commercial interests have also been quick to respond to the opportunities afforded by the Net, with British supermarket chain Sainsbury's offering virtual shopping, a tour round a typical store, and a home-delivery wine-ordering service.

Added to this is the deployment of food metaphorically in a whole host of cultural products. The dialogue from *Pulp Fiction* used at the start of this chapter is one current example, but food has long served as a useful symbol for conveying social, cultural and moral messages. Maggie Lane's *Jane Austen and Food* (1995), for example, tracks one novelist's culinary codes, while Cindy Dorfman (1992) shows how the kitchen has been used as the setting in recent US movies for passions more usually located in the bedroom. And as Beardsworth and Keil (1990) note, analyses of food consumption offer the opportunity to illustrate a whole host of social processes. Thus George Ritzer (1993) uses the notion of 'McDonaldisation' to describe changes in contemporary society; and Diane Barthel (1989), following the lead of Barthes' classic readings of 'everyday objects' in

Mythologies (1973), takes a long look at chocolates, chocaholics and chocolate boxes in a discussion of modern design and the modernist political programme. Kirsten Ross (1995) also considers modernity, this time through domestic technology such as the refrigerator. A central part of this book's project is to use food to think about space and identity, as will become clear later.

Academic interest in food consumption has, of course, been incredibly wide-ranging and prolific. Setting aside all the technical literatures, whether from health, nutrition and food science or from catering and the food business, we still have an enormous range of material – too enormous to do justice to. All we can do is provide a token reference or two in each area, which could serve as the starting point for further inquiry. In history, then, we have both general surveys (e.g. Levenstein 1988 on America; Mennell 1985 on England and France), work on specific histories (e.g. Schwartz 1986 on diets and food fads or Elias 1978 on manners) and studies of particular foodstuffs (e.g. Mintz 1985 on sugar; Visser 1986 on the components of 'dinner'). In sociology, Mennell, Murcott and van Otterloo (1992) have provided a very useful literature review which highlights the richness of sociological study into food and eating, while in neighbouring anthropology we have pioneering works by Lévi-Strauss (1964) and Mary Douglas (e.g. 1984), and many studies of particular food habits and traditions (e.g. Appadurai 1981 on Hindu South Asia). Archaeologists have also tracked foodways through time, often in league with anthropologists (e.g. R. Willis 1990). Psychologists have sought to understand the mental processes involved in our relationships with food, and have given particular attention to so-called 'eating disorders' (e.g. Brumberg 1988), which have also been approached – from sometimes very different angles – by medical sociology (e.g. Turner 1992) and women's studies (e.g. Bordo 1993; Charles and Kerr 1988). Other departures include international relations (e.g. Enloe 1990), cultural studies (e.g. S. Willis 1991), economics (e.g. Fine, Heasman and Wright 1996), communication and media studies (e.g. Fine and Leopold 1993 on food advertising; Sanjur 1982 on mass media's influence on dietary patterns), philosophy (e.g. Curtin and Heldke 1992; Telfer 1996), leisure studies (e.g. Kalka 1991), colonial and postcolonial studies (e.g. Burton 1993; Goldman 1992), work and organisation studies (e.g. Leidner 1993; Mars and Nicod 1984), technology studies (e.g. Cockburn and Ormrod 1993), social policy (e.g. Lang 1986/7), folklore studies (e.g. Lloyd 1981) and architecture (e.g. Langdon 1986). On top of these sometimes false divides, Brown and Mussell (1984: 13) emphasise the transdisciplinary and interdisciplinary approaches common in foodways research – which they describe as 'nexus studies', since food is 'posited as a nexus for the convergence of traditional disciplinary methods and insights'.

This brings us inevitably to geography, itself something of a nexus or hybrid discipline. Again, it is not proposed here to give anything approaching a comprehensive listing of available literature; mere pointers are all that can be given. If anything, the geographical literature on food has tended to be dominated by production and trade issues (e.g. Le Heron and Roche 1995; Tarrant 1985), which are largely outside the remit of this study (although it is also false to divide production from consumption, and far more useful to conceive of webs or networks integrating producers and consumers through space; see Fine and Leopold 1993). A special issue of the journal *Political Geography* (edited by Frances Ufkes in 1993) has explored the globalisation of agriculture, reflecting the interplay of circuits of consumption with those of production; texts such as that by Goodman, Sorj and Wilkinson (1987) also deal with agribusiness and biotechnology, stressing their spatial impacts. World patterns of food distribution, with themes such as surplus, hunger and aid, have also received considerable attention, notably, of course, from a 'development' perspective, but also by historical geographers (e.g. L. Young 1994, forthcoming). Kodras' (1992) discussion of the lengthening of the American breadlines is offered as a poignant 'geographical snapshot' of food politics in North America, as is Harvey's (1993) discussion of (the lack of) social justice in the 'broiler belt'.

Retailing has been a further focus of this political-economic perspective within geography – see a series of commentaries by Neil Wrigley (1987, 1989, 1991), for example. Social and cultural geographers have also become interested in shops and shopping, with recent work on malls being noteworthy (e.g. Chaney 1990; Shields 1989; see Jackson and Thrift 1995 for a review of 'geographies of consumption') although not without justified criticism for sometimes losing 'the shoppers and the shopped' in a fog of 'textual' interpretations of mall environments (see Gregson 1995). Elsewhere, geographers and others have written on 'workplace geographies of display' in waiting on tables (Crang 1994), on the importance of foodways in the maintenance of cultural (and place) identities (e.g. Brown and Mussell 1984), on historical geographies of dietary change (Grigg 1995), on the globalisation of food products (Arce and Marsden 1993; Cook 1994), on wine (Dickenson and Salt 1982; Unwin 1991) and beer (e.g. Rogerson 1986 on sorghum beer in South Africa; Sutherland, Williams and Mather 1986 on Taiwanese beerhouses), and on any number of the 'cultural landscapes of food', from Mexican restaurants in Tucson (Arreola 1983) to the whole range of ethnic restaurants across the USA (Zelinsky 1985), and from drinking-place names in the central USA (Steiner 1986) to the iconography of American gay bars (Weightman 1980). Geophagy – eating earth – has also (unsurprisingly) attracted geographers' attentions, with its rather literal linking of geography and 'food' consumption (e.g. Hunter 1993).

Most of the material mentioned here analyses aspects of food consumption *in itself*, although some of the authors discussed use food as an indicator of other social processes – drinking places in Taiwan are indicators of increasing Westernisation in Sutherland, Williams and Mather's (1986) article, for example. However, the geographical study of food consumption has potential to inform many other contemporary debates in social (including geographical) theory – those concerning globalisation, identities and points of identification, the cultural politics of consumption, so-called 'travelling theory', place politics, nomadism, and so on. Issues mapped in texts such as Waters' *Globalization* (1995) or King's collection *Culture, Globalization and the World-System* (1991) have clear resonances with current research into food consumption geographies, as work by the likes of Jon May (1993) and Ian Cook (1994) shows. In *The Condition of Postmodernity* (1989), David Harvey used food consumption as a clear example of the processes of 'time–space compression' he was theorising. The food market, he wrote,

> now looks very different from what it was twenty years ago. Kenyan haricot beans, Californian celery and avocados, North African potatoes, Canadian apples, and Chilean grapes all sit side by side in a British supermarket. This variety also makes for a proliferation of culinary styles, even among the relatively poor. Such styles have always migrated, of course, usually following the migrant streams of different groups before diffusing slowly through urban cultures.... But here, too, there has been an acceleration, because culinary styles have moved faster than the immigration streams. It did not take a large French immigration to the United States to send the croissant rapidly spreading across America, nor did it take a large immigration of Americans to Europe to bring fast-food hamburgers to nearly all medium-sized European cities.
>
> Harvey 1989: 300

For Harvey, then, the globalising forces of time–space compression have replaced the diffusion of culinary cultures via migration of ethnic groups with processes governed by worldwide media and capital flows, which target market niches, seize new opportunities, and can thus 'assemble the world's cuisine in one place in almost exactly the same way that the world's geographical complexity is nightly reduced to a series of images on a static television screen' (300); his use of TV as an analogy reinforces his comments on the separation of image from reality in contemporary society, often discussed using Jean Baudrillard's concept of the *simulacrum* – an image, or sign, which has come to replace the reality it signifies. Under the condition of postmodernity, then,

> the interweaving of simulacra in daily life brings together different worlds (of commodities) in the same space and time. But it does so in such a way as to conceal almost perfectly any trace of origin, of the labour processes that produced them, or of the social relations implicated in their production.
>
> Harvey 1989: 300

While these last criticisms have themselves become ways to find new market segments, through the rise of so-called 'ethical consumerism', and while we might want to signal a note of caution around what can sound like elitist judgments about mass (food) culture, Harvey's brief discussion of food consumption provides a useful example of how to think a key contemporary social process (time–space compression) through food.

Other current debates about geographical concepts such as 'home', or 'place', or 'space', or 'the local', together with renewed interest in issues such as nationalism, migrancy and in what is usually named 'spatial theory' generally (e.g. Bird *et al.* 1993; Carter, Donald and Squires 1993; Chambers 1994; Keith and Pile 1993; Robertson *et al.* 1994b; Shurmer-Smith and Hannam 1994), also benefit from thinking through food. Doreen Massey (1995), for example, recently discussed themes of place-identities in past and present – on her recent holiday, she says, 'the *real* France' was to be found in cafes with 'the smell of Gauloises, the taste of good coffee and croissants' (182) – but she then acknowledges that

> [t]he 'real' France which we breathe in at the cafe … is itself composed of influences, contacts and connections which, over time, have settled into each other, moulded each other, produced something new … but which we now think of as old, as established … the essential France.
>
> Massey 1995: 183

The past becomes sedimented into the present, and Massey argues that concerns with conservation and tradition deny the dynamic forces which come together as a 'conjunction of many histories and many spaces … in this place, now' (191). Food which signifies that essential France is specific to the present conjunction; cigarettes, coffee and bread, even in the local brandings and varieties that echo 'Frenchness', each have histories which have brought them together, now, to evoke France. In the seventeenth century, for example, London was as renowned as Paris for its coffee-houses (Mennell 1985), and we no longer think of coffee as any more quintessentially French than it is, for example, Italian; the migration of croissants to the USA, remarked upon by David Harvey, meanwhile suggests a future loss of their essential Frenchness as they become globalised. As Jonathan Friedman (1995) argues, current emphasis on cultural mixing, or *creolisation*, within the debates around globalisation, can serve to deny that *all* cultures are hybrid; his example of pasta having been introduced to Italy from China by Marco Polo shows this clearly, although he adds that 'for most [Italians], the Chinese connection is today quite irrelevant for the cultural definition of spaghetti' (Friedman 1995: 83). The cultural mix which brought pasta to Italy is lost, with spaghetti becoming quintessentially Italian. In the same way, while American fast foods, that staple derided example of the anti-local used by Massey, are presently un-French, the processes of creolisation could create, in the future, an equally disconnected cultural (re)definition of Le Big Mac, or the Royale with Cheese.

Indeed, McDonald's (and its Big Mac) has served the cause of cultural comment nobly, as both metaphor and 'text' for explicating different theoretical stances; its corruption into every neologism from 'McWorld' (global commodity culture) to 'McJobs' (low-pay, low-status service sector work taken by 'slackers', as described famously in Douglas Coupland's 1991 novel *Generation X*), from McDonaldisation (Ritzer 1993) to Parker and Jary's (1995) comment on changes in higher education, which lead to the 'McUniversity' (very distinct from McDonald's training centre, the 'Hamburger University'), gives it a truly omnipresent iconography, its golden arches marking landscapes from Tokyo to ... Paris. And in the world of social theory, too, McDonald's has iconic status. Staying with *Pulp Fiction* a second longer, we echo Jules Winnfield's request of Vincent Vega: 'Example.'

In a paper called 'Travelling theory/nomadic theorizing', Nick Perry (1995) offers an analysis of the Big Mac; his aim is to suggest that through 'nomadic' theorising – theorising which is 'disrespectful of boundaries and resistant to categorization' and is thus 'hybridized, mongrelized, customized, made promiscuous, invested with voice; not local, not lost but rather found elsewhere, in places where conventional theory does not (and cannot) travel' (43) – we can gain a fuller understanding of globalisation, including the globalisation of knowledge. From the perspective of organisation studies, McDonald's franchising scheme is often cited as exemplifying 'flexible accumulation at the expense of flexible specialization' (44), offering pseudo-entrepreneurship under conditions of extremely tight control. However, Perry is critical of this and certain other kinds of 'McTheorising' as we might call it – especially the kind deployed by the likes of Sharon Zukin (1991) – which subsumes cultural meaning under social and/or economic relations. Instead, Perry advocates a closer and more beefy reading of 'hamburger as text'. The Big Mac is, he writes, 'a conceptual approximation of the perfect commodity' (47), to the extent that it is used as a pseudo-currency in *The Economist*'s 'purchasing power parity' index, which measures the relative costs of buying a Big Mac across the globe. But the Big Mac is a *sign* as well as a commodity – a point reflected in Barry Smart's (1994) description of the varieties of (legal and not-so-legal) economic activity surrounding service at Moscow McDonald's. Perry writes that these two approaches (hamburger-as-text and hamburger-as-sign) 'move at the intersection between burger-nomics and burgerology' (49), or between differentiation and *differance*, to use a different theoretical parlance. In summary, Perry is suggesting that different ways of thinking (through food) can elaborate different theoretical perspectives – and, most fruitfully of all, the *spaces between* different perspectives can open up still newer ways of thinking.

THINKING FOOD/SCALING PLACES

It is part of the project of this book to approach food consumption and social and cultural theory in the kind of way Perry advocates. By deploying different theoretical perspectives, *and* reflecting on the spaces between them, we hope to open up new ways of thinking through food. As a structuring device, we have chosen the discussion of spatial scale contained in Neil Smith's paper 'Homeless/global: scaling places' (1993). By moving through the scaling *body–home–community–city–region–nation–global* (while acknowledging its limitations – see, for example, Woodhead 1995), exploring different aspects of food consumption at each scale, we aim to produce a textured account of what Cook and Crang (1996) call 'circuits of culinary culture' as they map across space.

Smith's manifesto for a scaled spatial politics begins with three stories – of the Homeless Vehicle designed by Krzysztof Wodiczko, of the riot in Tompkins Square Park, New York City, in August 1988, and of (spatial) struggles extending from there over the entire Lower East Side. These stories foreground his exploration of the spatialised politics of scale; his theoretical agenda also maps 'spatialised discourses', arguing for greater understanding of the social construction of space, which has, he says, been somewhat taken for granted within social theory: 'scale is produced in and through societal activity which, in turn, produces and is produced by geographical structures of social interaction' (Smith 1993: 97). In this way, the notion of scale can be seen as 'the geographical resolution of contradictory processes of competition and co-operation', with scale being 'the criterion of difference not so much between places as between different *kinds* of places' (99; emphasis in original). Echoing Perry's focus on the spaces between, Smith calls attention to the jumping of scales – the crossing of or resistance to boundaries. Scales are actively socially connected, not rigidly separate, and their schematic representation in a hierarchy is intended not as some 'ontological system', but as a reflection of the fact that geographical scale is 'hierarchically produced as part of the social and cultural, economic and political landscapes of contemporary capitalism and patriarchy' (101). Although Smith's emphasis is on spatial(ised) politics, a fact reflected in his choice of examples to illustrate each scale, the structuring of space via scale provides a useful way of ordering our discussion of regimes of food consumption, since the geographies of consumption at each scale (as well as between scales) provoke different theoretical and empirical responses.

The first scale in Smith's scheme, the body, has largely lain outside the geographical agenda; recent work by Jon Binnie (1996), Vera Chouinard and Ali Grant (1995), Julia Cream (1995), Robyn Longhurst (1995) and Gillian Rose (1993), however, has begun to situate the body in space. In particular, theories of the body's social construction and of the projection of identity through (or on to) the body suggest that we need to seriously consider the body within social (and spatial) theory (e.g. Shilling 1993). And, of course, food consumption is about the body, both at the physiological level – we all need food – and in terms of the ways in which we think about the relationships between the food we eat and our bodies; issues of health and illness, of purity and pollution, and, most profoundly in contemporary Western societies, of body weight, shape and image, mark the scale of the body as centrally important to unpacking food consumption geographies.

Extensive work, notably by feminist scholars, has focused on so-called 'eating disorders'. A series of essays by Susan Bordo, collected in *Unbearable Weight* (1993), charts women's food ambivalences in contemporary Western societies, expressed for some in 'disorders' such as anorexia or bulimia, and by a startlingly large proportion of all women in an endless cycle of diets, food fears and calorie counting. In response to women's troubled relationship with food, diet aids, health programmes and therapies of every sort are offered. Reviewing three forms of 'eating therapy' – NutriSystem, feminist psychoanalytic therapy and Overeaters Anonymous – Catherine Hopwood (1995) uses a Foucauldian–feminist perspective to expose the way each therapy pushes 'the subject in different directions' (79). Considering them in that order, Hopwood summarises the way each positions the compulsive eater, revealing very different constructions of the place of food in women's lives, and very different discourses of empowerment:

[T]he compulsion to eat is understood as: a behaviour to be controlled; a symptom of split-off part(s) of the self; or, a 'gift' to facilitate honesty. Eating itself may be: a controllable activity; a satisfying and pleasurable activity; or an indicator of abstinence and one's relation to a Higher Power. So, too, the relation to the body varies. It may be perceived as: an entity to be controlled; a fundamental and expressive aspect of the self; or, a temple to be honoured. Finally, relationships may be: transactions between separate persons; connexions vulnerable to symbiosis; or, responsibilities to be honoured.

Hopwood 1995: 79

Autobiographical writings by women often also consider food and ambivalence. Two collections, the 'feminist cookbook' *Turning the Tables* (O'Sullivan 1989) and *Eating Our Hearts Out* (Newman 1993), both contain material by women exploring their feelings about food. A passage by Susan Ardill, from *Turning the Tables*, encapsulates so many Western women's experiences of food and eating at the end of the twentieth century:

Sometimes food seems nothing but a problem; it consumes me, an endless source of pleasure and frustration, and indeed pain. Its endlessness is the problem – food is about boundaries, maps of the body, the outlines of social give and take. I wonder if I'll ever get into shape. Meanwhile, I love shopping for food; love finding and learning about new types of pasta, or cheeses; love the balancing act of cooking; and most of all, inexhaustively love eating.

Ardill 1989: 84

As Bordo (1993: 108) notes, it is women who habitually seek emotional heights from eating, but who are also 'most likely to be overwhelmed by their relationship with food, to find it dangerous and frightening'.

There are many other expressions of the relationship between food and the body, of course – from 'healthy eating' to 'heavy drinking', from food taboos to food allergies, from 'you are what you eat' to the way you eat it. For example, Munro's (1995) thesis on the necessary addition of *disposal* to theories of food choice and consumption jauntily considers the smell of old fish in the fridge, the policing of garlic breath and the effects of flatulence on sales of pulses (see also Elias 1978 on table manners).

The next scale, the home, is both a physical structure (for most people) and a site of social reproduction, in and around which routine acts – 'eating, sleeping, sex, cleansing, child-rearing' (Smith 1993: 104) – take place. The home is often defined as quintessentially 'private' space, the space of the family (often problematically – on exclusion and its resistance, see Colomina 1992; Johnston and Valentine 1995). Within geography, feminist work has again been central in unpacking the scale of the home, especially in exploring the gender divisions of labour therein, including the impacts of domestic technologies (e.g. Miller 1983). Kirsten Ross's account of French domestic life, in *Fast Cars, Clean Bodies* (1995), discusses the 'Taylorising' or time management of domestic work, at least in part a product of increasing domestic technology, and Lupton and Abbott Miller (1992) cast a Foucauldian eye over both bathroom and kitchen – and the conduct of housework – as forms of both docility and discipline. Research into the impacts of domestic technology on the division of labour in the home consistently reports that gender divisions remain stable, with women still doing most of the 'housework' (including food shopping, preparing and so on), while 'labour-saving' inventions have not impacted upon the amount of time spent on domestic work. Cynthia Cockburn and Susan Ormrod's (1993) work on microwave ovens, of example, echoes Thrall's (1982) conclusions: technology has a conservative effect on social relations within households, although microwaves were seen to increase children's involvement in food preparation. Family arrangements around food in Charles and Kerr's (1988) study of 200 women with small children showed a clear gendering at all levels, from choice of 'menu' to the size of portions. In the space of the home, the domestic politics of food bear more than anything else the marks of gender inequality.

The scale of the community is one which has attracted considerable attention from geographers and sociologists. For the former, it has been to some extent problematic, since it cannot be 'mapped' in any straightforward ways: it is more a 'structure of feeling' than a territory, although it does often have a shared, commonsense boundary (revealed when people are asked to delineate the extent of their own community). Of course, its use extends far beyond the immediate space around the home – what might better be called the neighbourhood – since different people have a different sense of belonging to different communities: a non-space-bound community, like the 'lesbian and gay community', for example, is very different from what people might refer to as their 'local community'. As Smith (1993: 105) puts it, community is 'the least spatially defined of spatial scales'; what is almost universal, though, is the 'vague yet generally affirmative nurturing meaning' attached to 'community'. Benedict Anderson's (1983) famous discussion of 'imagined community', and recent work on the 'virtual communities' of cyberspace (e.g. Batty and Barr 1994), echo Smith's point. In terms of food consumption, community is most often thought of in one of two ways. In the literature on migration and foodways, food habits are seen as a fundamental way of shoring up a sense of (usually ethnic) community identity (e.g. Brown and Mussell 1984), while material on local, place-bound communities (or neighbourhoods) considers food as social glue – Hunt and Satterlee (1986), for example, discuss pubs as social centres for communities; corner shops (often romanticised by anti-mall writings) are another good example of community consumption centres, as are ephemeral 'community events' like barbecues and street parties. Work by David Crouch (1989, 1991) on allotments, although the intertwining of themes of production (home growing) and leisure (or work, depending on your perspective) with those of consumption (eating home-grown fruit and veg) are also resonant here:

> The fun of working an allotment is that you can do it alongside other people. This brings trials when neighbours are too friendly to let you get on with your digging, but there are many pay offs. Sharing seeds, plants and produce go hand in hand with swapping hints and suggestions. The whole idea of allotment gardening is bound up with a community of shared ideals, if often contested opinion about how you should grow things.
>
> Crouch 1991: 38

Themes of sharing and co-operation, of 'common ideals' (recycling, organic gardening, the superiority of home-grown produce, the joy of gardening), thus mark the allotment as an exemplary form of 'community' in Smith's 'affirmative' sense. Even on the plot, of course, there are tensions, neighbour disputes (over encroachment, for example), competitiveness, factions, even sabotage and intimidation (especially where plotholders are 'growing for showing' in the summer round of competitions and shows). Class antagonisms, sexism, racism, homophobia – all these are features of what Crouch therefore quite accurately describes as 'an expression of local culture' (38), as surely as they crop up in the pubs, shops and 'local festivals' that embody food consumption practices at the scale of the community.

At the scale of the urban, we are in 'the most accomplished centralization of capital and social resources devoted to social production, consumption and administration' (Smith 1993: 107). What does this mean for food consumption? It means a concentration of choice – the contemporary Western city has an astonishing range of food shops, cafes and restaurants, take-aways, pubs and clubs, fast-food stands, home and office delivery services. Style magazine *The Face* recently ran a piece on Pizza Hut home delivery; their journalists and photographer went out on pizza runs, quizzing customers about why they were staying in, eating pizza, on a Friday and Saturday night in London. At 10.20 pm on Friday, 24 March, they delivered two small vegetarian pizzas to Simon and James, at James' new 'bachelor' flat in Islington:

So are you ordering to pizza to celebrate [moving into the flat]?

James: I suppose you could say pizza marks a rite of passage for me, ha ha. Of course it's also something simple to eat when you're busy.

No, stick with the cultural studies. This is good.

James: Well . . . what pizza says is that you're busy but you still want to entertain, to have people round. I like the idea of having stuff delivered to your door. It's very modern . . .

What's the best thing about staying in with a pizza?

James: You can save your excitement on a Saturday night. The friends and the conversation we have are more important than the pizza itself. But I appreciate the fact that it's an option . . . there isn't enough time in the day for everything but that hasn't got anything to do with pizza. Has it?

Benson, Thomas and Constantine 1995: 133

Not having enough time in the day, and liking the 'modern' home delivery of fast food are quintessential features of urban living; twenty-four-hour cafes and super-markets, specialist food shops, and the range of eating places – these mark the urban eating experience as poles apart from corner shops and allotments (although there are many of both within cities). They also separate the urban from the rural, where food provision is often severely underdeveloped (though, of course, the rural is commodified through food consumption in 'farm shops', quaint village tearooms and 'local', characterful pubs, all ripe for city day-trippers to pick). The current ebbs and flows of urban growth and no-growth, counterurbanisation and gentrification continue to reshape the map of the urban, but its food geography remains one of choice, convenience and access.

If the urban is about consumption, then perhaps more than any scale, the region is about food production. As Smith (1993: 108) says, 'the regional scale is closely bound up with the larger rhythms of the national and global economy, and regional identity is constructed disproportionately around the kinds of work performed there'. The region can be conceived as 'a concentrated network of economic connections between producers, suppliers, distributors and myriad ancillary activities, all located in specific urban or rural locations'. Even in these times of 'post-Fordism', of new spatial divisions of production and labour, the region clings to an identity forged in an era of productive – often mono-industrial – activity. Actions to protect regions, to shore up and reaffirm their identities in the face of this erosion increasingly nostalgise the region chauvinistically, although Smith notes that 'global anti-imperialist' projects represent a more progressive politicisation of the region.

A very clear example of regional politicisation around food appears in Warren Moran's (1993) paper 'Rural space as intellectual property', which discusses *appellations d'origine* and European Community regulation of the wine industry. *Appellations* function as a kind of trademark, allowing wine producers who satisfy certain production criteria to use a regional identity on their wines; for the customer, *appellations* signify certain standards of quality, often linked with artisanal principles of 'craftsmanship' in production. Within processes of globalisation, such geographical indications of provenance gain increasing currency (as discussed below), countering standardisation-through-internationalisation. Importantly, *appellations* mix the natural environment of the region with the 'raw materials' (grapes) used and the skill involved in production and processing, thus ensuring a tie to place. Mutual publicity thus occurs – wines are famous for coming from a particular region, the region is renowned for its wines – making wine regions and vineyard tours popular with tourists.

Moran (1993: 269) suggests that the idea of 'distinctive local and regional products is embedded in some national and regional cultures'. Thus, particular foods (cheeses are a good example), drinks, even whole cuisines, are attached to places. (Italian cooking is heavily divided by region, as is Chinese cuisine; see Lo (1994). For an attempt to regionalise English cuisine, see Ayrton (1980).) New World winemakers seeking to label their wines with *appellation* names have met resistance; while 'the US view is that the term Californian Chablis is appropriate … European wine law insists that such names describe wines made in particular regions of Europe and should be used only for wines originating from them' (272). Thus the region's character, including that of its producers, is literally bottled, and since wine-tasting remains a significant test and mark of distinction (in Bourdieu's sense), the market is always thirsty for a glass of *authentic* France, with some drinkers claiming they can tell the taste of particular vineyards, the year of production and even, as a recent British newspaper advert for a Microsoft CD-ROM wine guide jokingly suggested, the individual who made it.

At the scale of the region, elements of a chauvinistic 'regionalism' are often manifest. And, of course, the discourse of nationalism similarly echoes through the scale of the nation. Although there are many other dimensions to this scale – the power of the state is given prominence by Smith (1993) – nationalism has become an important focus for those wishing to understand the ideologies that map across space. And national identity is expressed through food consumption. For Japanese society, rice embodies the nation (Ohnuki-Tierney 1993); and we all regularly choose to eat Italian, or Chinese, or Indian (although, of course, what we eat is usually a hybrid tempered by our own tastes). Processes of migration, as already mentioned, transmit food habits across and between cultures; at times, where cultures clash, nationalism is both mobilised by and can mobilise food issues. In a paper on the changing social position of Turks in Germany, Ayse Caglar (1995) shows how the Turkish-introduced fast-food doner kebab (*kebap* in German) stands centre-stage in the renegotiation of a migrant group – especially in the context of the emerging *Ausländerfrage* ('question of foreigners'). Where the doner was initially marketed by Turks as an exotic ethnic food (bought overwhelmingly by Germans), and was used as a positive symbol of cultural cross-fertilisation in multi-culturalist discourses, the effects of changing attitudes towards 'foreigners' have led to efforts to downplay the Turkishness of doner, with stands and chains using names like McKebap and Donerburger, reflecting the McDonaldisation of German tastes. At the same time, 'doner' became a short-hand name for Turks – a multicultural youth festival in Berlin in 1987 was called *Disco döner* – and this was then incorporated into discourses around the *Ausländerfrage*, with Turks using slogans like *Kein döner ohne Ausländer!* (no doner without foreigners) in defence of their increasingly threatened place in German society. (It is noteworthy that within this political turmoil, doner kebabs are selling better than ever in Germany.) But for Turks, a continued association with doner means a continued denial of full upward social mobility; it is a final irony that, in attempting to further distance themselves from what they now see as a negative association, former doner sellers are moving into the restaurant business, catering Italian food.

As Caglar (1995: 223) notes, 'the transformations of *döner kebap* manifest the new ways of articulation and negotiation between the local and the global'. And Turkish entrepreneurs selling Italian cuisine to German customers bring us on to Smith's final scale, the global. Globalisation has become a site of considerable theoretical debate within the social sciences in 'postmodern' times: Featherstone and Lash (1995: 1–2) note that 'the global begins to replace the nation-state as the decisive framework for social life'. As Robertson (1995: 27) says, the twin processes of homogenisation and heterogenisation have 'become features of life across much of the late-twentieth-century world' – which leads him on to a discussion of a neologism originally coined in the pop-business literature – *glocalisation*. With global media, transnational corporations and food transport, processing and marketing systems spanning the world in seamless webs, food consumption might be said to be glocalising; that is, to be succumbing to the twin forces of universalism (McDonald's again) and particularism (the authentic exotic) (for a discussion of homogenisation and heterogenisation 'myths', see Miller 1995). Perhaps more than any other process, globalisation (or glocalisation) is shaping and reshaping patterns of food consumption. In the words of Curtis and Pajaczkowska (1994: 207),

> Gastronomic participation in cultural difference takes place along a spectrum that moves from the familiar to the exotic. If the experience of difference creates anxiety, then this can be compensated by a quest for a food that is as commonplace as possible: the friendly safety of finding chips, a recognized brand of beer or Coca-Cola; or contributing to the global success of McDonald's with its slightly inflected but predictable range of food offered in proximity to tourist attractions in cities throughout the world. If the experience of the familiar breeds contempt, alimentary adventuring may become part of the project of ingesting foreign culture.

For some of us, a familiar meal in an unfamiliar setting helps fight off the panics of disorientation. But for many other mouths, the excitement of 'eating the Other' is an important way of thinking about one's place in the world. We do not know whether Vincent Vega ingested foreign culture on his trip to France; if the thought of 'that shit' – mayo – on fries made him baulk, then we doubt it . . .

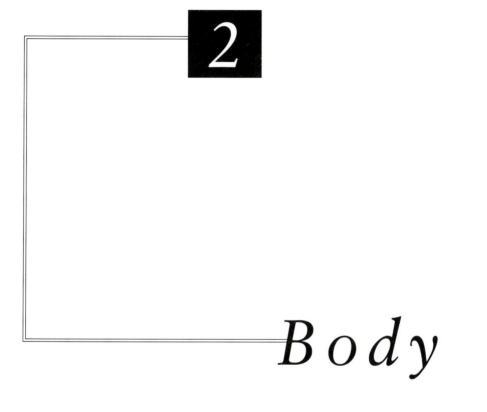

2

Body

MICHAEL DEAR (18)
Wittgenstein, Food and the Paradox of Creativity

●

You cannot write anything about yourself that is more truthful than you yourself are. This is the difference between writing about yourself and writing about external objects. You write about yourself from your own height. You don't stand on stilts or on a ladder but on your bare feet.

Ludwig Wittgenstein

The burgeoning subdiscipline of gastronomic geography owes much to Ludwig Wittgenstein, who was one of the first to recognise that we are what we eat. As a centurion, I am frequently asked (by admirer and detractor alike) about the link between my diet and my intellectual stature. Standing barefoot in front of the mirror every morning is a harrowing experience for most of us, but I comfort myself by anticipating what next awaits me: a hearty and wholesome breakfast! Most of my scholarly *bons mots* explode into consciousness as I dutifully masticate my morning bran. I confess that this sometimes upsets the dog, but we all agree (as a family) that the occasional mess is a small price to pay for a paradigm shift. I hope you agree . . .

OATMEAL

For each serving, take the following ingredients: ⅓ cup of 1-minute oatmeal; ⅓ cup of milk (whole or 2 per cent); ⅓ cup of water. Combine ingredients in a pan and bring to a boil; reduce heat and simmer for about one minute, or until mixture reaches consistency of a fumarole. Pour into dish. Now comes the *pièce de résistance*. Take a 1-lb can of Tate and Lyle's Golden Syrup (imported if necessary); add syrup to the oatmeal, stirring continuously, until the mixture can absorb no more.

Note bene: on no account should salt be substituted for syrup (look what happened to Scotland). And if you try to use maple syrup, well, just forget it (look what happened to Canada).

Serve your cerebral ambrosia immediately, garnished with the Comics section of the morning newspaper and a cup of hot tea (with milk but, please, no sugar). Let the sweet nectar glaze its way through your gullet, percolate promiscuously around your pericardium. Then shiver with anticipation as the sensuous slither of scholarly insight seduces your synapses!

BODY

Eating is an important bodily practice and indeed is one of the most effective ways we can shape and remake the space of our bodies. This chapter uses the medium of food to explore geographies of the body. It begins by exploring the relationship between what we eat and the amount of space we take up. We outline geographically and historically specific social norms of the 'ideal body' against which we all measure our own. Drawing on Foucault's understandings of how the body has historically been subdued, we go on to explore the social and cultural practices of discipline, surveillance and self-restraint which shape our bodies in everyday spaces, such as the home and the workplace. We also consider geographies of bodies that deviate from the 'ideal' norm: the fat body and the anorexic body.

The chapter then goes on to consider the body as cultural capital, exploring how patterns of consumption and practices of the self such as eating and table manners mark class differences on to the body. This theme of 'we are what we eat' continues in the next section of the chapter when we consider some of the ways in which we manage our bodily boundaries in order to avoid eating foods that are contaminating and polluting. In particular, we focus on cultural classifications of food as edible/inedible and good/bad for our health.

Finally, we think about some of the ways in which food has been used as a metaphor for the body itself, exploring the relationships between appetites for food and sex, food as a sexual commodity and the emergence of gastro-porn.

2

BODY

•

Our exploration of food, place and identity begins 'not with a continent or a country or a house, but with the geography closest in – the body' (Rich 1986: 212). The body, Adrienne Rich argues, is a *place of location*. Describing how her lived experiences are inscribed or mapped on to her body, she writes:

> I see scars, disfigurements, discolorations, damage, losses, as well as what pleases me. Bones well nourished from the placenta; the teeth of a middle class person seen by a dentist twice a year from childhood. White skin, marked and scarred by three pregnancies, an elected sterilization, progressive arthritis, four joint operations, calcium deposits, no rapes, no abortions, long hours at a typewriter – my own, not in a typing pool – and so forth. . . . [T]o locate myself in my body means more than understanding what it has meant to me to have a vulva and clitoris and uterus and breasts. It means recognising this white skin, the places it has taken me, the places it has not let me go.
>
> Rich 1986: 215

In other words, it means recognising the *politics* of this location. This chapter uses the medium of food to explore the politics of the body – examining the relationship between food and the space our bodies take up; the role of eating in the production, and disciplining of bodies; and the processes by which we manage our body boundaries.

TAKING UP SPACE: FOOD AND BODY SHAPES

The British broadsheet newspaper *The Sunday Times* regularly carries a column entitled 'Relationship of the Week' in its supplement. Among those relationships featured has been that between the Duchess of York and her weight. Branded the 'Duchess of Pork' by the tabloid press, Fergie is one of many celebrities, including Italian opera singer Pavarotti, US chat show host Oprah Winfrey and English footballer 'Gazza' (Paul Gascoigne), who have come under scrutiny by the media for taking up too much space (Plate 2.1). In contrast, the bodies of the Princess of Wales and international supermodels such as Jodie Kidd and Kate Moss have been the subject of media attention for being too thin. The watch company Omega even went so far as to withdraw its advertising campaign from the fashion magazine *Vogue* (31 May 1996) in protest at what it described as the 'anorexic' models used by the magazines in fashion shoots (Figure 2.1). These examples highlight the centrality of food and eating, first, to the social representation of the body, and second, to the social production of the body (Turner 1984).

The body is far from being a natural phenomenon. Discourses in the media, the fashion industry, medicine and consumer culture map our bodily needs, pleasures, possibilities and limitations. These cartographies produce geographically and historically specific social 'norms' within which we each locate, evaluate and understand our bodies (Gamman and Makinen 1994). The foundation of modern consumer culture, thanks to the emergence in the 1920s of a new media of mass-circulation magazines, tabloid newspapers, the cinema and radio, proved particularly important in establishing and promoting new bodily norms. Advertising began to create a new discourse of self-improvement centred around needs and desires, in which images of beautiful bodies were used to entice people to buy the new consumer durables on offer. Body maintenance itself even became a marketable commodity in this consumer culture (Featherstone 1991). Now in the late twentieth century we are bombarded with visual images of ideal bodies against which we can measure our own. Featherstone (1991: 178) writes, 'Images invite comparison: they are constant reminders of what we are and might with effort yet become.'

While studies in the USA and UK suggest that modern Western nations are getting fatter, ideal body shapes are getting thinner (Garner *et al.* 1980; Stephens, Hill and Hanson 1994). The weights of Miss America Pageant contestants, *Playboy* centrefolds and fashion models have all declined over the past three decades (Garner *et al.* 1980; Morris, Cooper and Cooper 1989; Stephens, Hill and Hanson 1994). This trend is evident in the editorial content of women's magazines (Guillen and Barr 1994). Two studies, the first of magazines published between 1969 and 1979 (Garner *et al.* 1980) and the second of magazines published between 1959 and 1988 (Wiseman, Gray and Mosimann 1992), have traced a significant increase in the amount of space devoted to dieting and slimming. A further study of *Ladies Home Journal* and *Vogue* has mapped a corresponding change in the desired female body shape promoted by these magazines, away from a curvaceous figure towards a thinner, more androgynous appearance (Silverstein *et al.* 1986). Even advertisements for food products in women's magazines

Plate 2.1 Celebrities' body shapes are under constant public scrutiny
Photograph: David Maddison

Figure 2.1 Omega caused a publicity storm by banning 'anorexic' models from its ads
Reproduced from the *Guardian*, 31 May 1996, by kind permission of David Austin

appear to place more emphasis on dieting than eating (Klassen, Walker and Cassel 1990–1).

Bombarded as they are by these images, it is not surprising that many women are dissatisfied with their own bodies – skipping meals and feeling guilty about eating. A survey by the US magazine *Glamor* found that of the 33,000 women who responded to its questionnaire, 75 per cent claimed that they were fat even though 30 per cent of them were actually medically underweight (Bordo 1993). Even women who are happy with what they weigh often feel dissatisfied with their body shape or particular parts of the body, such as the thighs or the hips (Charles and Kerr 1986a). When researchers carrying out a study at the Melpomone Institute for Women's Health offered participants a range of body shapes and asked them to pick their ideal body, 44 per cent chose a shape that was 10 per cent underweight (Stephens, Hill and Hanson 1994): 'Few women accept their bodies as they are' (Charles and Kerr 1986a: 543). A woman interviewed in a study by Charles and Kerr explains:

> 'I can't stand being overweight and if I do go much over 9 stone it starts to show round my hips. John is terrible. He says 'look at you' ... and that's it, I start dieting because it's so noticeable. I can't bear to think that he can see me looking fat.'
>
> Interviewee quoted in Charles and Kerr 1986a: 545

Although women in particular are subject to popular discourses eliding representations of slim bodies with health and sexual attractiveness in what Lupton (1996: 137) calls the 'food/health/beauty triplex', men's bodies are also increasingly the subject of similar discourses evident in magazines and television advertisements (Mort 1988; Bocock 1993; Jackson 1994); a recent issue of the British *Men's Health Magazine* (June 1996) offering men advice on 'how to drop 10 lbs fast', and the Australian public health advertisement 'Gutbusters' (Lupton 1996) are just two examples of the contemporary discourses circulating around the masculine body. Like the ideal feminine body, the ideal masculine body is expected to be fat free. But it is also increasingly being sexualised by the use of muscular, eroticised images of men in advertisements and magazines (Dyer 1982; Mort 1988). As Bocock points out,

> Men – gays, bisexuals and 'straights' – are now as much a part of modern consumerism as women. Their construction of a sense of who they are, of their identity as men, is now achieved as much through style of dress and body care, image, the right 'look', as women's.
>
> Bocock 1993: 105

In case we should forget about the body shapes we should all be aspiring to, tabloid newspapers and popular magazines are always on hand with 'horror' stories about gross bodies that take up too much space – men who can't fit into airline seats and women so fat they block out the sun on the beach. In this contemporary cultural context, where a fat body is understood to be unhealthy, ugly and sexually unattractive, the emphasis is on individuals to be self-disciplined in order to manage or produce their bodies in a culturally desirable way. Bordo (1993) highlights three different ways of thinking about the relationship between the body and the mind which she traces back to the philosophers Descartes, Augustine and Plato: first, the body as inseparable from the mind but also simultaneously different (other) from it – a stranger; second, the body as the cage or prison of the mind, which the mind seeks to escape; and third, the body as an enemy, distracting the mind from useful tasks with its constant lusts and desires for both food and sex – the 'sins of the flesh' (see, for example, Cosman 1976; Lawrence 1988; Lupton 1996).

This third understanding – that the mind has to struggle to keep the body in line by fighting its demands for food – is particularly prevalent in discourses on dieting where metaphors of battle, such as 'fight the flab', are used to urge potential consumers to buy diet foods and exercise equipment and sign up for gyms and slimming clubs (Bordo 1993). The slim body represents the mastery of the mind over the body which threatens to get out of control.

It is the French philosopher Michel Foucault whose work has been most influential in shaping our understanding of how the body has historically been subdued. His book *Discipline and Punish* (1977) opens with a graphic description of a Frenchman, Damiens, being tortured to death in 1757. From this beginning, in which Foucault demonstrates how the vengeance of the sovereign was literally acted out on the body of the eighteenth-century prisoner, he goes on to trace the development of penology, illustrating how this direct intervention on the subject was replaced by the detailed organisation and control of prisoners through social space. Bentham's Panopticon – a circular prison giving the warders perfect visibility into every cell – established surveillance as a mechanism of disciplinary power without any physical instrument other than architecture and geometry. This technology has proved particularly potent precisely because in Foucault's (1977: 206) words it gave 'power of mind over mind' so that individuals would exercise self-surveillance and self-discipline. He writes:

> [T]here is no need for arms, physical violence, material constraints. Just a gaze. An inspecting gaze, a gaze which each individual under its weight will end by interiorising to the point that he is his own overseer, each individual thus exercising surveillance over and against himself.
>
> Foucault 1977: 155

In this way subjected bodies – what Foucault (1977: 138) termed 'docile' bodies – are produced, which serve prevailing relations of domination and subordination.

These social and cultural practices of discipline, surveillance and self-restraint take place not only within institutions such as prisons and school, but also within the everyday places to which are bodies are oriented (Fiske 1993). At different geographical locations – at home, school and work – we learn how to present, maintain and reproduce our bodies over space and time according to what Turner (1984: 190) terms 'dense systems of social norms and regulations'. Bordo (1993) in particular stresses the 'direct grip' that culture has on our bodies. She writes:

> [T]hrough routine, habitual activity, our bodies learn what is 'inner' and what is 'outer', which gestures are forbidden, and which required, how violable and how inviolable are the boundaries of our bodies, how much space around the body may be claimed.
>
> Bordo 1993: 16

In a service-based economy, the bodies of employees have become a crucial part of the work process, affecting the outcome of transactions. Consequently, employers are putting increasing emphasis on managing the bodily performances of their interactive service workers (McDowell 1995). As Leidner explains, 'Interactive jobs make use of their workers' looks, personalities, and emotions, as well as their physical and intellectual capacities, sometimes forcing them to manipulate their identities more self-consciously than do workers in other kinds of jobs' (Leidner 1991: 155–6). In a study of two organisations, McDonald's and Combined Insurance, Leidner (1993) exposes the lengths that these employers go to to discipline their workers' bodies – specifying exactly how workers should look, what their body shape should be, how they should dress and how they should speak and think. Even the layout of fast-food restaurants, Leidner argues, is designed to promote the constant surveillance of employees. In a study of merchant bankers in the City of London, McDowell (1995) documents how similar disciplinary tactics at the bank have produced a striking physical uniformity among the male staff; while Anna (Box 2.1) describes the role everyday conversations about dieting and calorie-counting played in promoting self-surveillance and bodily discipline among her former colleagues in a British accountancy firm.

The home is often assumed to be a private place – a haven from the 'public' pressures of the world of work. Saunders (1989: 184) describes the home as a place 'where people are offstage, free from surveillance, in control of their immediate environment'. But as geographers have increasingly begun to recognise, this division between public and private is overstated. The privacy *of* a location is not the same as privacy *in* a location. The distinction between 'public' and 'private' is difficult to draw within households when individual boundaries are blurred by shared activities such as cooking and eating, and when there is little or no space where household members may have privacy from each other (Allen and Crow 1989). Rather than a place where we are, as Saunders claimed, 'free from surveillance', the home is actually a place where many of us are subject to the

Box 2.1 ANNA

ANNA NOW WORKS FOR AN ENVIRONMENTAL CHARITY IN A DERBYSHIRE VILLAGE, BUT USED TO WORK AS AN ACCOUNTANT IN A CITY-CENTRE OFFICE.

'One thing I noticed when I was an accountant, every other female in the office was on a diet [laughs]. It used to really annoy me. They would talk about it endlessly and most of them were not overweight at all. It seemed to be the whole culture of the place that if you were female you had to be on a diet. And luckily the place I'm at now, everybody tucks into their chips and cheese pie at lunchtime, it's just not the same at all.

'Ninety per cent of the [accountancy] office was paranoid about dieting. Well in my view paranoid. They probably thought it was perfectly normal. I think I was the only woman who was not on a diet, and even a large proportion of the men were on diets. I had one manager who went on the Hay diet, that was sort of food combining.

'Ahm, there was one woman who was a little overweight. Not markedly so, ahm, but as she was always trying to lose weight, I think it was Weightwatchers, you know, she got weighed every week and had certain numbers of units of various foods. Ahm, and she would tend to explain to people exactly what she was doing as well. But most of the rest, particularly the women, tended to turn down cakes and things like that – "Oh, I can't possibly have that" – and have cottage cheese on their sandwiches for dinner.

'I think they were all sort of image conscious. Ahm, it was the sort of area where you get the young professionals who have to wear the right suits, and I think part of that image consciousness is about weight rather than about health. I think they weren't really that bothered about what they ate; it was more the number of calories in it. Which to me seemed a bit daft, but that was they way they seemed to feel about it.

'You see, it was just so irrational. I mean, these were all sensible, educated people, but they could never actually see that dieting might have an irrational side of it. I mean none of them ever thought about a healthier lifestyle generally, or like taking more exercise. It was all about going on the latest diet. Which to me is absurd. And these are people who probably knew a fair amount about nutrition. But they still went along with all this.'

surveillant – often very intimate – gaze of partners, children, parents, relatives and close friends. For some this means the home is more important than the workplace as a site for disciplining the body. It is when we are naked in the bedroom and bathroom that all our wobbles and bulges are most commonly on display. Women, in particular, complain of marital and family pressures to lose weight and attain an 'ideal' body shape (McKie, Wood and Gregory 1993). Tracey Fewston told the British tabloid newspaper the *Daily Mirror* (5 July 1996: 15) that she lost 7 stone (98 lb) after her husband told her she looked like a beached whale. The same paper also ran a double-page spread entitled 'I'm leaving you because you're fat' (*Daily Mirror* 14 August 1996: 6–7), revealing the stories of women who were dumped by their husbands because of their body shape but went on to shed pounds and find 'true' love again.

Without the disapproving gaze of loved ones, those who live alone, like Jackie (Box 2.2), claim they struggle to exercise self-discipline, sometimes letting their bodies slip out of control. The home with shelves, cupboards and a fridge packed with edible items and all the technology at hand to produce instant gratification is a site of private temptation, so much so that women attending a workshop on food, described by Lawrence (1988) in her book *The Anorexic Experience*, claimed they rarely kept their favourite foods in the house so that temptation could not get the better of them.

Box 2.2 JACKIE

JACKIE IS 43 AND A LONE PARENT. SHE HAS TWO ADULT DAUGHTERS AND AN 11-YEAR-OLD SON.

'I mean, you sit at home on your own at night and you think "what can I nibble?" and you try and eat something that isn't fattening but it doesn't always work ... I mean, you know sometimes when you're sort of sat in the house on your own in the evening, like the kids are all in bed ... and you're sort of sat there and whatever's on telly's probably not that interesting and you think, "Oh this is boring, there's nobody to talk to, I'm fed up, what can I do? I know, I'll go and have something to eat" ... When you're just cooking for yourself as well it gets a bit of a bore 'cos you think, "What shall I do tonight", you know, and it's just not worth bothering sort of doing a lot of stuff sometimes for yourself. I start looking at packets more.

'I mean we [lone parents] constantly talk about, you know, things to do with men and dieting and everything else, you know, mainly because a lot of us haven't got a man in our lives.... I reckon if I got a man I'd probably lose weight, that's a good incentive to lose weight ... because you want to look nice for them, don't you, to some extent. Even if you call them names and everything you still want to look, if you've got a fellow, you still want to look nice if you're like going out with him or whatever. It's definitely an incentive.

'Oh God, my mother used to feed me like she was feeding the bloody army. She's a fantastic cook my mother.... But you know, I mean, a big dinner plate when I was a child, I would get a big dinner plate piled high with food and I would eat it so it's probably my mother's fault that I got as fat as I did.... I used to eat these enormous meals – I mean they were absolutely beautiful ... you know, loads of mashed potato and meat or stew or whatever she'd done and oh it was lovely – she can't half cook, my Mum. So I mean, I did, I ate very big meals, you know, which were far too big really for someone of my age.... I got really, really fat when I was sort of, well, beginning of me teens.'

The home can also be the site of fragmented bodily practices. Parenting is intrinsically about knowing, managing, controlling and caring about children's bodily activities. Teaching children table manners, toilet training them, restricting their play to keep them safe and disciplining them by smacking or grounding them are just a few of the numerously repeated and often unacknowledged ways that mothers and fathers exercise power over their children's bodies (Morgan 1996). But parents' own bodily practices do not always coincide with the rules they establish for their offspring. Often it is a case of 'do as I say, not as I do'. While many adults, especially women, diet, parents anxious about the dangers of eating disorders often prevent their teenagers from doing so (Bell and Valentine 1996). In contrast, adult household members usually have bodily licence to eat as they choose, especially if they cook for themselves. It is not uncommon for multiple meals to be on a household dinner table each evening (see Chapter 3): a 'proper meal' for the children; a low-calorie meal for one parent who is trying to lose weight; and a high-carbohydrate diet for the athletic parent who is trying to discipline their body in a different way. In this way food is one medium through which parents articulate power over their children's bodies, although of course some children will always try to resist this, for example by smuggling food they do not want out of the 'public' domestic space of the dining-room and flushing it down the toilet, or by bingeing on chocolate bars in the privacy of their bedroom or the 'public' space of the park.

Sometimes rather than disciplining the body, social practices in particular locations, like the home or the workplace, can actually have the opposite effect, driving the body out of control. Mike (Box 2.3) was happy with his body shape until he started working for a large manufacturer. His job as a buyer necessitated regular business meetings with sales staff that were always conducted over lunch or dinner. In a profession where a double chin and beer belly were a badge of corporate success, Mike was unable to avoid eating at work without losing his business social networks. Similarly, Jackie (Box 2.2) blames the origins of the problem she has controlling her weight on the huge home-cooked meals her mother fed her as a child.

As these two examples demonstrate, many of our family albums resemble 'before' and 'after' magazine pictures. It is not just at major stages in the life course – pregnancy and old age, for example – that we experience a change in body shape. For most of us, throughout our lives we experience our bodies as always 'in a process of becoming' (Shilling 1993) but never actually there; never actually in that 'ideal' shape.

PLACES A FAT BODY WON'T LET YOU GO

'The body is political, central to personal identity and life chances' (Synott 1993: 33). A slim, fit body is for some a source of pride to be paraded in public places, spelling

Box 2.3 MIKE

MIKE IS 42, A COLLEGE LECTURER WHO USED TO WORK AS A MARKETING MANAGER.

'I worked in buying and buying was, you were always taken out for lunch by sales people who came to visit you and, er, so you know, I don't know, possibly three or four times a week, really, you'd be taken out for meals and go to restaurants or pubs or whatever and, er, oh, you know, you'd go away and you'd be entertained when you were away and so that was like eating a great big meal at lunchtime and then eating a big meal in the evening to be sociable as much as anything.

'I mean, five pints at lunchtime was common, you know, in buying; that, you know, really, that was dead common and people used to drink more than that, I mean, some people might have eight, ten pints ... the buying, the sales people in buying you know, the sales and buying interface was very much that, sort of, you know, male get-up at a pub, a pub and you drink, so you have sort of two or three pints before your meal and then a pint with your meal and then a pint after and that was, yeah, that was a business lunch, so it was very much built around food and, yeah, you know, it was important, you know, thinking about it, in terms of it was the relationships that you made there and if you couldn't eat or couldn't drink you'd be in real, you wouldn't be able to keep the job really, people would have looked on you as being really strange, you know that, well, you know, "I don't want someone who can't eat" ... if someone said, "no, I'm sorry, I don't want to do that", you were sort of frowned on, pushed to one side, you had to be a "hail fellow well met" type of person.

'I started to put, well I started to put a lot of weight on because I was just eating all the time, I remember coming up to Christmas and I'd been out for 18 or 19 Christmas dinners ...

'I mean, it was basically my stomach that, that started, er, I just got, you know, a big stomach and, er, my face was fat ... but no, certainly, it was my stomach that, er, that just, you know, put weight on so the things like, you know, trousers, you, so I had to, you know, buy new, you know, big pairs of trousers and things like that.

'I started doing things to, to cope with the number of calories that I was putting in me, rather than doing things 'cos I enjoyed the sport – and that's when I started cycling to work really – so it was that stage when, you know, it was business lunches and all these sorts of things were, were, er, making, you know, a much, having a much bigger effect on my size and everything and then from cycling I started to, er, get interested in running and, er, you know, gradually that took on a much more, more important part, such that, you know, the tail wagged the dog really, you know, the running and the cycling became more important than, er, you know, keeping my body in some sort of shape.'

discipline, success and conformity, whereas fat is seen as a sign of moral and physical decay. Fat people are stereotyped as undisciplined, self-indulgent, unhealthy, lazy, untrustworthy, unwilling and non-conforming (Dejong 1980; Dejong and Kleck 1986), as the British tabloid newspaper the *Daily Mirror* (14 June 1996: 5) headline 'Fatties are thick says snooty mag boss Alex' illustrates.

Unlike the disciplined slim body, the fat body is not welcome in many everyday places. The Disney Corporation and the supermarket chain Asda are two organisations which appear to have their own corporate bodily norms to which aspiring employees must conform (McDowell 1995). Candidates who are significantly overweight stand little chance of being hired. McDowell quotes an employer interviewed as part of her study of a City of London workplace, describing how one potential merchant banker was rejected because he was too fat. He told her, 'We had a very nice chap in for interview but he was very overweight ... we sat and talked about it very seriously, about whether the fact that he was very large was going to weigh on the clients' mind' (interviewee quoted in McDowell 1995: 83).

Those with disciplined, docile bodies who are recruited by major organisations are often expected to maintain their weight and size as a condition of their employment (McDowell 1995). The New York City Traffic Department is one of many famous institutions to have fired employees for becoming overweight (Schwartz 1986).

Children can be equally discriminating. Mathews and Westie (1966) asked 144 US high school children to rank a range of photographs according to statements from 'I would be willing to marry this person' to 'I would exclude this type of person from my school'. Their results revealed that 'not one student wanted to share their personal, emotional or geographical space with a fat child' (Cline 1990: 233).

As Cline (1990) points out, fatness is a visible stigma: you can't pass as slim (Plate 2.2). Everyday places and facilities are planned for specific body shapes. 'Airplane seats, subway turnstiles, steering wheels in cars are designed to make fat people uncomfortable. People in motion in the modern world should be as streamlined as their vehicles' (Schwartz 1986: 328). Schwartz's frustration is shared by Cline's interviewee Trudie, who says, 'I ... get angry on buses when I notice everybody passes me by. Nobody wants to sit next to me. I take up too much space. I ought to be angry with public transport, but I get angry inside' (interviewee quoted in Cline 1990: 227).

Faced with these sorts of experiences, Foucauldian self-surveillance can inhibit the everyday geographies of fat bodies. The beach is one particular location where the body is expected to be more visible than usual. Shields (1991) argues that the carnivalesque zone of the beach liberates bodies from the disciplinary practices of the workplace and everyday codes of moral conduct. The beach, 'with its structure of the natural, legitimates ... sexual display' (Fiske 1989: 52). But for those whose bodies do not live up to the exacting standards of the food/health/beauty triplex, the beach is a place of embarrassment, self-

Don't Be Too Fat

Don't ruin your stomach with a lot of useless drugs and patent medicines. Send to Prof. F. J. Kellogg, 1866 W. Main St., Battle Creek, Michigan, for a free trial package of a treatment that will reduce your weight to normal without diet or drugs. The treatment is perfectly safe, natural and scientific. It takes off the big stomach, gives the heart freedom, enables the lungs to expand naturally, and you will feel a hundred times better the first day you try this wonderful home treatment.

Plate 2.2 The fat body
Photograph: Corbis-Bettmann

consciousness and even self-concealment. Cline quotes the experience of 12-year-old Jannie, who says:

> 'It's worse on the beach. Mum's so ashamed of my size she carries this huge towel to shroud me like a corpse. She makes me hold it round me all the way to the water, then she calls out: "You can drop it now, Jannie, run and hide in the waves".'
>
> (interviewee quoted in Cline 1990: 226)

In response to fat discrimination in the USA, the National Association to Aid Fat Americans held a Fat-In in New York in 1967. The idea that power relations are articulated through the body and that it could therefore be a site of opposition and resistance was also an important aspect of second-wave feminism in the 1960s and 1970s. Writers such as Brownmiller (1984) argued that the trappings of femininity – corsets, high heels and so on – served to discipline women's bodies to reproduce patriarchy. Women were encouraged to all give up all forms of body maintenance, from dieting to shaving their legs, because these bodily practices were seen to be about pleasing men. Fat was, as Orbach (1988) famously claimed, a feminist issue. Ironically, these strategies of resistance in turn became disciplining practices, producing their own 'docile' – in this case, unmaintained – bodies (Green 1991). In the late 1980s and 1990s there has been a backlash against this regulatory framework as feminists have sought to re-engage with femininity on new terms.

Now information technology offers us all the possibility to forget body politics by liberating our minds from their bodily prisons. Walter Hudson, the world's fattest man, according to *The Guinness Book of Records*, was so large before his death in 1991 that he was unable to go out, and spent his days immobilised in bed surrounded by all the essentials for existence in a high-tech society: a fridge, a toilet, a television and a computer. His PC enabled him to live a 'disembodied existence', shrinking his body to whatever shape or size he desired to become a virtual traveller courtesy of computer-mediated communications (Morse 1994). For Hudson and many others, Morse (1994: 179) suggests, one of the great seductions of this technology is that 'the apparatus of virtual reality could solve the problems of the organic body, at least temporarily, by *hiding* it.'

MINIMISING SPACE: ANOREXIA

Given the prevalence of popular discourses about the ideal body and fat discrimination, it is not surprising that some people take the issue of controlling their body's cravings for food to the extreme where they seek to kill off its demands altogether – achieving a complete ontological split between the mind and the body. Jenefer Shute, in her fictional account of a woman's battle with anorexia, *Life Size*, describes how the main character, Josie, strives to subdue her body and minimise the space she takes up while sneering at those she believes greedily occupy too much space, having given into their bodies' appetites. Josie sees her flesh as superfluous, masking her pure, authentic self:

> She [her nurse] says, I am a starving organism and my brain is starving and therefore not working the way it should! On the contrary, it's never been purer and less cluttered, concentrated on essentials instead of distracted by a body clamouring for attention, demanding that its endless appetites be appeased.... It's not my fault if everyone around here is always stuffing themselves. This always made her [Josie's nurse] angry, which is why I said it. But it was clear to me that she was ... occupying so much space already, gobbling up much more than her share, so crassly exceeding bounds.... How could I justify the space I occupied in the world? Only by resolving to be stronger; by thinking of the next thing I would not eat, and the next, and of nothing else.
>
> Shute 1993: 7, 122–3 and 158.

The term anorexia was coined by the British doctor William Gull in 1873, although the first case was not formally identified in the USA until 1945. It was still a rare phenomenon in 1973 when Hilde Bruch published the first significant account of eating disorders; yet by 1984 it was estimated that between 1 in 200 and 1 in 250 US women aged 13–22 years old were anorexic (Bordo 1992), and by 1990 the American Anorexia and Bulimia Association was claiming that 150,000 US women were dying of anorexia every year (Wolf 1991). Bruch argued that anorexia is not about food and weight loss but rather about discipline and denial. Bordo (1992), for example, points out the synchronicity of the anorexia epidemic with other contemporary cultural practices involving control or mastery of the body, such as bodybuilding and jogging.

The feeling of control over the body gained when women start to diet gives some such a sense of satisfaction and achievement that some just keep on going. Starvation is merely the method for achieving control over the body, weight loss the outcome (McSween 1993). As Jenefer Shute's fictional character Josie found when she started to diet, she was admired by her school friends not so much for her size but for her self-control. Often, sticking to her starvation strategy involved having to duck the efforts of her parents and the school to provide meals for her, by managing the cartography of her eating, concealing and then disposing of food at home and at school.

When willpower fails, anorexics often feel guilty and punish themselves physically, as this anonymous woman describes:

'It was during the sixth form that I gradually realised that something was wrong. I began to cut down drastically on the amount I ate at meal times. . . . Week by week I would have no other aim than to have lost a pound or two. . . . Yet if there was no change I would feel wretched all week. Then my efforts to lose weight would become more desperate and every day I would punish my body with bouts of exercise. The worst feelings were the sense of guilt and self hatred which would come over me. . . . I stayed at school late to avoid eating with the family, and meals left for me in the oven were surreptitiously buried in the dustbin. My thinking was single minded – I had to control my weight at all costs.'

Anon. 1992: 126

Numerous explanations, from low self-esteem and family dysfunctions to teenage traumas, have been put forward to explain why it is predominantly middle-class women and girls who are affected by what Chernin (1992) has called 'the tyranny of slenderness' (see, for example, McSween 1993). Bordo (1992) offers two explanations. First, she argues that anorexia can articulate a fear and rejection of women's traditional gender roles. For these women it often starts at puberty and is an attempt to hold back the tide of development – to eliminate the breasts and prevent menstruation. This account is also favoured by other writers such as the psychiatrist Arthur Crisp (1974, 1980) and the feminist Kim Chernin. She writes:

[M]any women feel a 'terror of female development', of taking control of their own lives, and that eating disorders, which occur at times of underlying developmental crisis, regardless of biological age, work subconsciously to prevent movement into the public sphere.

Chernin 1983: 21

Second, Bordo (1992) argues that anorexia is a fear of 'the female'. She says the anorexic is terrified of

a certain archetypal image of the female: as hungering, voracious, all-needing, and all-wanting. It is this image that shapes and permeates her experience of her own hunger for food as insatiable and out-of-control, which makes her feel that if she takes just one more bite, she won't be able to stop.

Bordo 1992: 44

Chernin (1992: 67) quotes a woman she spoke to as saying:

'I have rarely had a moment of peace about my body. All my life, no matter what

else is going on, I have felt an uneasiness. A sense that something was about to
get out of control. That I needed to keep watch. That something about me was
changing.'

Interviewee quoted in Chernin 1992: 67

Compared to the volume of work published on women and eating disorders, there
has been little research which attempts to 'map the difference "race" onto the
complexities of gender and body shape' (Gamman and Makinen 1994: 157). While some
writers such as Gordon (1990) claim that eating disorders like bulimia are less common
among black and Asian women, suggesting that they are less prone to be influenced by
Western ideals of the slender body, others have argued that it is naive to believe that
because black women have largely been excluded from these representations, they are
free from the pressures on white women. Gamman and Makinen (1994) point out, for
example, that this very obliteration of black women's experiences of their own bodies
actually serves to construct 'beautiful' black women as those with light skin and
straightened hair.

Certainly there is some evidence to suggest that eating disorders are becoming more
common among non-white ethnic groups in both the UK and USA. Research by
Mumford, Whitehouse and Platts (1992) using an Eating Attitude Test found that British
Asian schoolgirls have more unhealthy attitudes to eating than white schoolgirls and that
British Asian girls have a higher rate of diagnosed eating disorders than their white
counterparts (Ahmad, Waller and Verduyn 1994). Explanations for this racialised
pattern of anorexia and bulimia are as diverse as those offered to account for the
gendered nature of this epidemic. Bhadrinath's (1990) work suggests that it may be an
outcome of the complex dietary restrictions and fasting that Muslims undergo during
Ramadan (although not all Asians are Muslim). Gillespie (1995: 201) claims that
'Punjabi culture provides legitimating opportunities for extreme dieting practices: fasting
for religious reasons is common among women and girls, who may fast up to two days
a week, especially when they are "petitioning God" for a special request'.

Ahmad, Waller and Verduyn (1994) point the finger of blame at racism for
undermining British Asian girls' self-esteem. Other researchers have linked it to problems
of managing hybrid identities, suggesting that if the parents put pressure on their
daughters to adopt traditional dress and behaviour rather than allowing them to make
their own choices, the girls may try to regain their lost sense of control by dieting (Rezek
and Leary 1991; Ahmad, Waller and Verduyn 1994).

What the anorexic, the fat and the docile body all show is that

People construct and use their bodies, though they do not use them in conditions
of their own choosing, and their constructions are overlaid with ideologies. But

these ideologies are not fixed; as they are reproduced in body techniques and practices, so they are modified.

Frank 1991: 47

THE BODY AS CULTURAL CAPITAL

Consumption represents 'the endless producing and reproducing of desire, of the body in the world's image and the world in the body's image' (Frank 1991: 63). Just as gendered patterns of consumption mark bodies in different ways, so class difference is also marked on the body.

In a famous study, titled *Distinction: a social critique of the judgement of taste*, the French sociologist Pierre Bourdieu argued that 'the body is the most indisputable materialization of class taste' (Bourdieu 1984: 190). Featherstone sums up his argument thus:

> Classes reproduce themselves by their members' internalization and display of certain tastes, which then mark only some for distinction. At the foundation of these tastes is the body. Taste is *embodied* being inscribed onto the body and made apparent in body size, volume, demeanour, ways of eating and drinking, walking, sitting, speaking, making gestures etc.
>
> Featherstone 1987: 123

Bourdieu (1984) described the knowledge we have about which foods to choose, which cutlery to use and how to look after our bodies as 'cultural capital', arguing that practices of the self, such as eating, betray people's origins or *habitus* (internalised form of class conditioning). In a survey of over 1,000 French citizens, conducted in the 1960s, Bourdieu found strong class distinctions in terms of what the respondents ate, their methods of preparing food, their manners, and their attitudes towards the body. One of the most famous and comprehensive pieces of work exploring cooking, cuisine and class is Goody's (1982) book of that title, which considers the reasons why elite and peasant cuisines develop in some societies and not others.

Research in the UK has found that educated professional people tend to eat a greater variety of food and are more likely to eat healthy foods that are high in dietary fibre, exotic fruit and vegetables, and brown bread, and are more likely to be vegetarian than lower-income groups whose consumption patterns are characterised by 'vulgar foods' – foods rich in fats, tinned vegetables and fruit, white bread, more sugar and full-fat milk (Calnan 1990; Mennell, Murcott and van Otterloo 1992). The pattern is similar in relation to the preparation and cooking of food: the educated classes are more likely to bake or grill their food, whereas lower classes fry it.

Choices about where to eat out and which dishes to eat also mark bodies in different ways. Food writer Andy Harris, who runs 'gastrotours' on the Greek island of Evvoia, says:

'In the Eighties people went trekking in Nepal. Now they are coming on cookery courses. . . . Just as in the Seventies it was considered avant-garde to talk openly about sex and in the Eighties it was actually thought interesting to brag about mortgages and obscure decorative finishes, in the Nineties the emphasis is on pushing back gastronomic limits.'

Quoted in Foulkes 1996: 42

Diane Simmonds (1990) describes this explosion of food snobbery – which she terms Foodie-ism – as being on a par with the fashion industry. As in other matters of taste, of course, those with high economic capital do not necessarily possess cultural capital!

Those in different social classes also develop different body images because of the way they exercise and maintain their bodies. Sport, beauty treatment and surgery reproduce embodied class inequalities. Turner (1992) claims that each class has a sport which demonstrates its economic and cultural capital: 'weight lifting articulates working class bodies, while jogging and tennis produce a body which is at ease in the middle class milieu or habitus' (Turner 1992: 88).

Gillespie (1995: 198) argues that taste is also 'one of the most significant markers of ethnicity in plural societies'. In a study of young British Asians she found that although the young people who took part in her study ate Indian food at home with their families, they avoided this food at school, preferring to eat traditional 'English' and Americanised fast foods. She argues that for these students, eating Western food is a way of exhibiting some control over their own bodies and of articulating their hybrid identities. She states:

Unlike hair and skin colour, 'ethnic signifiers' which are more or less completely beyond personal control, body shape can, to some extent, be controlled by means of dieting. Body shape ideals are strongly influenced by the media and may also have a 'racial', ideological dimension. 'Indian' food is regarded as being fattening, due to the amount of *ghee* added to curries.

Gillespie 1995: 200

She goes on to argue that Indian food is often referred to by the students as 'village food' and is compared unfavourably with 'Western food', which is seen as exhibiting more cultural capital, a distinction she likens to Bourdieu's (1984) classification of

French rural peasants' food as fattening and low status. In contrast, the students' parents attempt to impose a different hierarchy of tastes upon their children – in which social difference is marked out in terms of different staple Indian foods. Similarly, one participant in Lupton's (1994) study of childhood memories of food recalls how she refused to take the food eaten by her family – Eastern European immigrants to Australia – to school because it marked her out as 'different'. Rather, she chose to have that most Australian of foods, Vegemite, on her sandwiches.

THE OUTSIDE IN: YOU ARE WHAT YOU EAT

'Food is a liminal substance; it stands as a bridging substance between nature and culture, the human and the natural, the outside and the inside' (Atkinson 1983: 11 quoted in Lupton 1996: 16–17). The mouth – another liminal zone, being neither quite inside nor outside, but somewhere in between – is the route by which food crosses these boundaries (Falk 1994). It is the link between social and bodily expressions of control, being both an aperture of the body and a social entrance and exit (Levens 1995). Falk (1994) describes the mouth as a safety chamber positioned between three 'gates' that protect the self from taking in dangerous substances from the outside. The first are the basic cultural rules about edibility, which determine what is allowed into the mouth. The second is the irreversible decision to swallow and so incorporate the outside inside the body. The third is the intermediate notion of taste, which is linked to other bodily senses such as touch and smell but is also inextricably bound up with cultural meanings (Lupton 1996).

The cross-cultural anthropology literature on food shows the importance of culture in determining what we eat (Lévi-Strauss 1964/1970; Douglas 1966). Since symbolically, as the phrase goes, 'we are we what we eat', food 'must not only be good to eat but also good to think' (Lévi-Strauss 1964/1970). Before we consume something we must be able to recognise it as a foodstuff, identify it, understand its place in our world and classify it as edible. In every human society, from all the range of potentially edible, nutritional items available, only some are considered appropriate or fit to be eaten (James 1990). Mary Douglas (1966, 1979) for example, points out how dogs and foxes are not eaten in modern Western societies, despite being eaten in many other cultures; and there are many examples of foods which are eaten by some religious societies but classified as 'unclean', 'impure' or polluting by other faiths. She argues that these cultural variations in cuisine are a product of the way different societies order the universe and assign value and status to people, animals, plants and insects. Potential sources of food which are compatible with a culture's taxonomy will be classified pure and edible. Those that are not threaten members of that culture's ontological subjectivity because consuming the wrong food may pollute or transform the self from within. Fischler (1988) points out, for example, that cannibals are often associated with taking on the characteristics of the victims they devour.

Cultural definitions of edibility in terms of individual foodstuffs and the order and combination of foods are time and space specific. White gravies were very popular on the early nineteenth-century US dining-table, although they are conspicuously absent now (Weaver 1986). For some religious communities the day of the week still determines what it is appropriate or inappropriate to consume; and chronological time of day still profoundly influences our understanding of edibility – steamed pudding and custard are not palatable for breakfast. Place also matters. What is edible in terms of individual items and combinations of foods when you are at home alone is very different from what would be considered food fit to be served at a restaurant; and both regions and nations have their own cultural understandings of what constitutes a meal (as later chapters illustrate). Douglas (1979) argues that these classifications – the rules which govern the structuring and timing of meals, and the practices and representations associated with the production and consumption of food – are what determines a society's cuisine. Transgressions of these cultural norms are considered revolting, sometimes inducing the body to vomit in disgust at what it has consumed.

But we are not only symbolically what we eat, but also literally what we eat, in that the food we eat provides energy and maintains our biochemical systems. According to Fischler, all food has medical significance because it can have an effect on us, in that it is 'the first and probably main means of intervening in the body' (Fischler 1988: 280). Although the notion that food influences health can be traced back centuries, it was not until the late eighteenth century that nutritional science developed (Lupton 1996). In the nineteenth century it took off as a result of government concern about the inadequate (in terms of quantity and quality) diet and consequent poor growth and ill-health of working men and soldiers (Turner 1991).

Lupton (1996) claims that there have been three major stages in the subsequent development of nutritional science. First, in the early twentieth century vitamins were identified, named and linked to the prevention of particular medical conditions such as rickets. Consumers were urged to eat so-called 'protective foods' which are rich in vitamins such as milk, eggs and vegetables. Second, after the Second World War a greater emphasis was placed on methods of food production. Third, since the 1970s the relationship between food and bodily health has played an important part in debates about the relationship between nature and society. Economic progress has allowed those of us who live in modern Western societies to enjoy a lifestyle characterised by overconsumption and physical inactivity – a lifestyle contrasted unfavourably with human society in a more 'natural' state. In the past two decades a range of degenerative illnesses has been linked to contemporary diets (for example, a high-calorie diet is linked to an increased risk of breast cancer), provoking anxiety among governments in modern

Western societies about spiralling health costs and lost economic productivity as a result of this bodily neglect (Turner 1992). As a result there has been a shift away from individualistic conceptions of the relationship between food and health towards state surveillance and regulation of diet based on scientific research. Eating habits are now at the heart of the discourses on prevention, and individual responsibility for bodily well-being, which dominate contemporary Western health education (Backett 1992). Examples include Target Health in Sweden and the Minnesota Heart Health Project in the USA (Jansson 1995).

British government health campaigns, such as *Eating for Health* (Department of Health and Social Security 1978), have stressed not only the importance of food to children's and adults' well-being, but also the significance of food to the development of the foetus in the womb (Murcott 1982a; MacIntyre 1983). Recent medical reports suggest that caffeine can increase the risk of miscarriage and stunt foetal growth (*The Independent* 1993a), while eating foods rich in folic acid, such as broccoli, Brussels sprouts and spinach, can reduce the risk of having a baby with spina bifida (*The Independent* 1993b). There are now a number of books on the market aimed at providing nutritional advice for pregnant women, such as Swinney's (1993) *Eating Expectantly*. Bordo argues that these discourses, placing mothers as responsible for providing an appropriate 'environment' (regardless of what fathers and indeed the State are doing) for the beginnings of the body, have the effect of privileging the foetus over the mother, rendering her a non-subject.

Like the classification of food as edible or inedible, the categorisation of individual foodstuffs as 'good' or 'bad' for our health and physical development is also a product of our culture. In some societies these definitions are linked to scarcity, concepts of 'naturalness', or 'freshness', designations of 'hot' and 'cold' and what constitutes a 'proper meal'. They also vary widely according to gender, age, class, religion and lifestyle (Lupton 1996). For example, those who are part of the natural-health movement or who adopt what may be loosely described as 'alternative lifestyles' in particular value 'natural foods' and reject artificial foodstuffs and those that have been produced by chemically intensive modern farming practices. Historically, and certainly within post-war modern Western societies, scientific advice about the merits and dangers of particular foods has fluctuated wildly (Anderson 1986; MacIntyre 1983). Potatoes, once regarded as unhealthy and fattening, are now recognised as an important dietary staple. Other foods, such as meat, have managed to straddle these binary definitions, being understood as simultaneously 'good' and 'bad': meat is valued for its protein and iron content and has historically symbolised wealth, power, strength and virility (Fiddes 1991); but it has also been identified as a potential contributory cause of obesity and heart disease as well as being associated with contamination by bacteria, growth hormones and bovine spongiform encephalopathy (BSE) (Lupton 1996). Perhaps not surprisingly, then, participants in Calnan's (1990) study of food and health exhibited a degree of scepticism about the value of food health warnings.

The way adults teach children to eat plays as an important part in the production and reproduction of these food moralities (Mead 1980). 'Children are rewarded for eating conceptually "good" meat and vegetables by being given conceptually "bad" pudding as a treat to follow. The "bad" food – the treat – is withheld if the "good" is not eaten' (James 1990: 681). Ironically, these practices construct foods that are 'bad' for us as the most pleasurable (Mead 1980; James 1990). For example, parents often use sweets to reward children for taking unpleasant medicine or doing chores (Charles and Kerr 1988).

A study by Backett (1992) found that respondents often used the language of religion – guilt, conscience, indulgence, gratification and deprivation – to describe their eating behaviours, particularly to articulate the guilty pleasures of indulging in unhealthy, 'naughty but nice' things. These moral judgements about 'good' and 'bad' eating habits also slipped easily into judgements about 'good' and 'bad' individuals (Backett 1992). Somewhat paradoxically, however, the interviewees, while describing health as something to be achieved, also acknowledged that there was a difference between what they knew to be 'good' for them and their actual eating habits. Despite all the contemporary health promotion campaigns the consumption of confectionery has actually increased over recent decades. Research suggests that 95 per cent of the British population eat chocolate confectionery at least once a day (Keynote Report 1990 quoted in James 1990). While lying in a grey area on the boundary between food and non-food, expenditure on confectionery represents a larger market than that for many dietary staples, such as bread and milk (James 1990).

Backett's (1992) interviewees used geographical time–space constraints to rationalise this paradox, fatalistically arguing that they could not fit a healthy lifestyle into their everyday routines while juggling the demands of home and work. Women in particular complained that they did not have enough time to manage their own bodies because of their responsibilities for their children and the home. These problems are often exacerbated by the need to negotiate significant changes in eating practices between household members. It is hard for an individual alone to change their diet if the rest of the household is resistant to new foods appearing on the dinner table and puts pressure on them to 'give up'. Men from meat-eating, 'hot-food' households, for example, are often reluctant to switch to eating salads (Twigg 1983). But preparing different meals for different individuals is time-consuming, expensive, and can complicate or disrupt the social occasion of the shared meal (McKie, Wood and Gregory 1993). Households often negotiate these differences by offsetting one member's activities against another's. For example, one person's low-fat diet may be traded off against another's drinking.

Eating to protect the body against decay or enhance its performance is now commonplace among top athletes. In the foreword to a book titled *Eat to Win* (Haas 1985), former tennis star Martina Navratilova states:

> I have discovered the secret of achieving peak performance and endurance in competition and training. The seemingly boundless energy and incredible stamina that have helped me to win so many world tennis championships in recent years, including Wimbledon in the summer of 1984, spring from the special nutritional programme I have been following.
>
> Navratilova, quoted in Haas 1985: xi

Haas analyses the blood of his athletic clients to calculate exactly what foods they need to put into their bodies to achieve the optimum health and sports performance. Like many diets, his approach draws on notions of the body as machine – notions that ideal inputs and outputs can be measured and quantified, and bodily requirements calculated. As James writes,

> the twentieth-century body literally embodies ideas of technological progress. It has become machine like. Its size, shape and capabilities are no longer taken as fixed, its limitations given. Rather, the body is seen as a potentiality that can be altered, improved and mended.
>
> James 1990: 667

These discourses about using body-work dietary technologies to perfect body-shape are particularly evident in male body-building cultures and advertisements for dietary supplements in men's magazines (White and Gillett 1994). While women are obsessed with bodily aesthetics, men appear to be equally concerned with the functional qualities of their bodies (Jones 1993).

TAKING IN THE WORLD: BODILY POLLUTANTS

'By taking food into the body, we take in the world' (Bakhtin 1984: 281); we incorporate it into ourselves and it becomes part of us (Fischler 1988). Food is supposed to be comforting but paradoxically it is also experienced as invasive, intrusive and contaminating. As Lupton (1996: 114) argues, 'to take in food – any kind of food – is to risk the integrity of the self by threatening pollution'.

Unlike many animals, which can get all the nutrients they need from one food source (the koala bear, for example), humans need to eat a diverse range of foods to obtain all the proteins, carbohydrates and minerals we need. This creates what Fischler (1988) terms the 'omnivore's paradox'. On the one hand, because humans, as omnivores, need to have a varied diet to survive we are inclined towards being innovative and experimental in what we eat. On the other hand, we have to be very wary or 'conservative' about what we consume because any unknown food is a potential source of danger. Fischler (1988: 278) explains, 'The omnivore's paradox lies in the tension, the oscillation between the two poles of *neophobia* (prudence, fear of the unknown, resistance to change) and *neophilia* (the tendency to explore, the need for change, novelty, variety).'

The preparation and cooking of food plays an important part in resolving this paradox by aiding the transformation of foodstuffs from Nature to Culture. As Fischler explains:

> [O]nce cooked [and] adapted to the conventional rules of a particular cuisine, food is marked with a stamp, labelled, recognized – in a word identified. 'Raw' food is fraught with danger, a 'wilderness' that is tamed by culinary treatment. Once marked in this way, it is seen as less dangerous. It can safely take its place on the plate and then in the eater's body.
>
> Fischler 1988: 287

But for some people 'marking' food in this way is not enough to tame it and make it safe. For those who suffer from allergies, the banal act of eating is fraught with potentially dangerous consequences. Some people need to control everything that comes inside their bodies. Not being able to eat wheat, barley and rye without encountering a severe allergic reaction which can include diarrhoea, vomiting, exhaustion, skin complaints and so on profoundly limits the everyday geographies of coeliac sufferers. As many supermarkets stock only a handful of suitable products, the simple act of shopping for basic foodstuffs can become a lengthy chore for a coeliac, involving time-consuming journeys to multiple retail outlets. International travel is limited by the need to choose the destination, travel company and airline to ensure that 'safe' food will be available. Even day trips in the UK or a meal out at a restaurant can be a logistical feat. The geography of an individual's social network can also be transformed by the coeliac diagnosis. Exchanging food and drink is a taken-for-granted part of many friendships. Having to constantly explain the condition and check the content of any foods offered by friends, neighbours and relatives is embarrassing, and many coeliacs find that their social networks gradually slip from being focused around local friends to being centred around others suffering from the same allergic condition.

The problems coeliacs encounter tracking down 'safe foods' among the hundreds of products on offer on supermarket shelves is being exacerbated because food production is becoming more remote from consumers, with most of the preparation of foods taking place not in the kitchen, but in industrial units. The food which reaches our table is often so processed and presented that we have little idea about its origins and history. As Fischler argues:

> [F]ood technology is becoming increasingly powerful in the sense that it now uses more and more sophisticated processes tending to mask, imitate and transform 'natural' or 'traditional products'; reconstituted proteins, artificial flavours, preserving techniques, etc. Quite literally, we know less and less what we are really eating.
>
> Fischler 1988: 289

Once new food technologies were seen as the way to turn raw food that was riddled with dangers (bacteria, putrefaction, etc.) into edible products; now to many people it is the mysterious and hidden additives, flavourings and production processes, rather than 'natural' raw foods, which represent danger (Fischler 1980). Coeliacs in particular are very sensitive to and therefore aware of the subtle and often unpublicised ways manufacturers continually alter production processes and the contents of everyday foodstuffs.

But it is not only the allergic body which is anxious about its vulnerability to invasion by polluting foods; Lupton (1996: 114) argues that 'the modern self is obsessed with notions of hygiene, purity and personal cleanliness'. She claims that ever since bacteria were discovered (in the Victorian era), we have been haunted by the invisible enemy – germs (see also Corbin 1986). As Sibley writes, 'Maintaining the purity of the self, defending the boundaries of the inner body, can be seen as a continuous battle against residues' (Sibley 1991: 3). One of the main themes of Douglas's (1970) work on the body as a classificatory system was the human response to dirt – which she famously defined as 'matter out of place' (Douglas 1970: 48). She argued that 'Dirt, and its equivalents, pose a cultural problem. It constitutes an anomaly. It stands out against a culturally prescribed pattern and will not fit in' (Murcott 1993a: 130) – a metaphorical threat of disorder which she extended from marginal matter to people.

Murcott (1993a) has used some of Douglas's concepts to explore bodily margins, particularly the danger and disorder posed by matter – urine, faeces, vomit – which is emitted from them. Using the example of the way we talk about babies and food in relation to adults and food, her work demonstrates the fundamental importance of body management in our everyday lives. She argues that babies are socially constructed as pure and innocent beings who must be protected (literally and metaphorically) from the pollution that their own bodies can create. Whereas adults can deal with their own bodily wastes and can regulate their own eating by evaluating their hunger, thirst, indigestion, constipation, diarrhoea and vomiting, parents have to monitor babies' diets by constantly checking the colour and consistency of their faeces (Murcott 1993a). As adults we learn that defecation must be separated in time and space from cooking and eating and that other bodily emissions, such as farting and belching, should be contained – at the very least – to the privacy of the home. To do these things in the wrong place or at the wrong time is to risk 'spoiling our identity' (Goffman 1964). But babies and young children have no concept of the importance of separating bodily functions. As a result, the mothers interviewed by Murcott (1993a) explained that they found themselves perpetually cleaning and sterilising clothes, utensils and the home in order to protect the babies from themselves, but also to protect other members of the household from the threat of the babies' bodily pollutions. By the 'law of contagion' we fear that when two things come into contact they transmit something of their essence from one to the other, so that foods which have come from or been touched by something or somebody considered 'offensive' threaten to contaminate us (Rozin 1987).

In the late 1980s and early 1990s the British population has gone from a state of ambivalence about food safety to one of collective anxiety following intense media coverage of a range of food scares (Mitchell and Greatorex 1990). First came the story that salmonella in chickens was contaminating British eggs; then listeria was found in cook-chill foods and cheeses; more recently the panic has switched to beef, which, allegedly, may be polluted with BSE. In between, food terrorists have tried to hold manufacturers and supermarkets to ransom by contaminating foods with glass and antifreeze (Beardsworth 1990). These uncertainties have had dramatic short-term effects on patterns of consumption, as well as causing long-term shifts in shopping and eating habits – for example, organic foods have established a firm foothold in the UK market (James 1993).

In an attempt to allay public fears, the British government has sought both to tighten up regulations covering the production, transportation and storage of foodstuffs, and to provide advice about hygiene in domestic practices. The space of the home has become an important site in the battle to maintain bodily boundaries against contamination. As Lupton and Abbott Miller argue,

[a]s settings for physical sustenance and hygienic care, the kitchen and the bathroom – and the product 'worlds' they frame – are crucial to intimate bodily experience, helping to form the individual's sense of cleanliness and filth, taste and distaste, pleasure and shame. These rooms are the home's most heavily invested 'objects' of domestic labour: failure to meet the high standards of hygienic maintenance attached to them is a source of guilt and embarrassment.

Lupton and Abbott Miller 1992: 504

The development of modern consumer culture has promoted a preoccupation with domestic cleanliness and waste. Ross (1995: 98) claims that the refrigerator – an object of desire in the 1950s – was designed to convey 'the image of absolute cleanliness and new-found hygiene: its brilliant white finish was the physical embodiment of health and purity'. Canned foods too were marketed not only as being convenient but also as being 'purer' than other foods because they eliminated both the mess and the necessity, with all its inherent risks, of handling foods (Dorfman 1992). But a whole host of consumer durables, developed with the stated intention of saving time and labour around the home – the vacuum cleaner, the washing machine, floor polish, oven cleaners and so on – have actually served to establish more rigorous standards of domestic cleanliness and to expand the bodily and domestic grooming and cleaning regimes that householders, especially women, are expected to undertake. The home in this sense has, like the body, become a 'disciplinary' site in which the modernised consumer kitchen epitomises Foucault's concept of docility (Lupton and Abbott Miller 1992). When feminists established a camp at the Greenham Common nuclear base the British tabloid press focused on the women's dirty cooking utensils and the disorderly way they cooked – an image they then contrasted with the model of cleanliness and order which is supposed to characterise the kitchen (Plate 2.3) of the heterosexual 'family' home (Cresswell 1996).

It is not only polluted, contaminated or unhygienic foods that can symbolise danger or upset our digestion. Any food – through its taste, texture, smell or appearance – can evoke disgust and aversion. Visser (1991: 311) suggests that 'We hate whatever oozes, slithers and wobbles' – a claim which Lupton justifies by arguing that

> [s]ubstances of such consistency are too redolent of bodily fluids deemed polluting, such as saliva, semen, faeces, pus, phlegm and vomit. Such bodily fluids create anxiety because of the threat they pose to self-integrity and autonomy. Body fluids threaten to engulf, to defile; they are difficult to be rid of, they seep and infiltrate.
>
> Lupton 1996: 114

Many contemporary artists have played upon these anxieties, outlandishly using foods to simulate bodily fluids. Karen Finley, for example, used chocolate to give the appearance of having smeared her body with excrement (Morse 1994).

Animal rights activists have also attempted similar shock tactics, using images of dead flesh and bloody carcasses or pictures of cute animals with captions such as 'how can you eat anything with a face?' in campaigns and advertisements promoting vegetarianism. Recognising meat as the flesh of a once-living creature or witnessing distressing incidents such as the killing or dismemberment of animals at a farm or the butchers is a common motivation for giving up meat (Beardsworth and Keil 1992).

Plate 2.3 Cleanliness is next to godliness
Photograph: Corbis-Bettmann

FOOD: EROTIC POSSIBILITIES BETWEEN BODIES

While food can arouse disgust and repulsion, it can also be erotic and pleasurable. A number of writers have used food as a metaphor for parts of the body; D.H. Lawrence, for example, used figs as a metaphor for the vagina, and the photographer David Thorpe has produced a number of books interweaving parts of women's bodies and foods, titled *Rude Foods* (see MacClancy 1992). Many different cultures also make linguistic links between eating and sex (Lévi-Strauss 1964/70; MacClancy 1992). Foodstuff names, such as honey, sugar and sweetie, are used as terms of endearment in modern Western societies; many body parts, like the breasts and penis, have food nicknames like melons and salami; while 'eat me' and 'suck my lollipop' are just two of the euphemisms for oral sex. The mouth is a particularly important site of erotic desire. Biting, sucking, licking and chewing are all bedroom – as well as dining-room – practices: one of the sexual fantasies described by a woman in a study on bulimia by Gamman and Makinen (1994) was to have chocolate licked off her body by the England football team in front of a cheering Wembley crowd.

Foodstuffs can also be sensual, evoking emotions and memories. In Laura Esquivel's Mexican magic realist novel *Like Water for Chocolate* (1993), which was made into a film of the same name, the heroine, Tita, has the ability to project her emotional state into the exotic foods she cooks. Her dishes then miraculously have the power to induce Tita's emotions in those who consume them. Tita uses this gift to communicate her desires, her passions and her jealousies, with both comic and tragic consequences.

Advertisements for various food products often play upon foods' erotic possibilities (Plate 2.4). Perhaps the most famous are the Häagen-Dazs posters featuring naked bodies smeared suggestively with ice cream. The Cadbury's commercial in which a woman lying in a bath longingly sucks a phallically shaped Flake also makes the most of food as a metaphor for sex – although, as Gamman and Makinen (1994) point out, this advertisement elides the woman and the food such that she is not only sexualised but also commodified and presented as consumable too. Male fantasies about consuming women are not limited to advertising images; several movies, including *Silence of the Lambs* and *Cape Fear*, represent women being devoured in whole or part; and the Australian band Kool And The Gang told one journalist that their fantasy was to place a woman on a platter with fruit and vegetables and then baste her like a turkey (Vardy 1996).

The idea that foodstuffs can promote the assumption that women's bodies are commodities for men's sexual consumption is a theme in Vardy's (1996) paper on sweet-scented, food-imaged toys. She argues that toys such as the Strawberry Shortcake series of dolls, Tonka's Cupcakes and Mattel's Popcorn Pretties, which took the US market by storm in the 1980s, 'carry with them a great deal of ideological baggage, which is all too easily assimilated into play and into children's views of the world' (Vardy 1996: 275). In

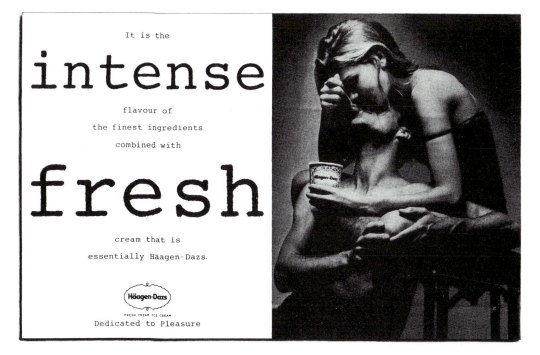

Plate 2.4 Häagen-Dazs markets the erotic possibilities of ice cream
Source: *The Sub*, March 1992, Newspaper Publishers Association

a detailed and sophisticated account of these products she also shows how the toys articulate the assumption that it is women's role to service men, not only sexually, but also by feeding them and providing other forms of domestic and moral support for them – a theme that we will return to in the following chapter on the home.

Coward (1984) suggests that gourmet features and advertisements in women's magazines have created a food pornography that seduces women in the same way that conventional pornography tempts men. This connection between food and sex is also shared by Margaret Reynolds (1990), who included the recipes of cookery writer Elizabeth David in her edited collection of women's erotica (Gamman and Makinen 1994)! 'Real porn too is being seasoned with food', according to Foulkes (1996: 42), who cites the opening chapter of Linda Jaivin's novel *Eat Me* as an example of gastro-porn.

The appetite for food has historically been linked to sexual appetite. In medieval times both were considered to represent animalistic desires and a lack of self-discipline. 'To give in to gluttony was considered as opening the floodgates for other sins and vices, to allow the devil within the body' (Lupton 1996: 132). Both temptations of the flesh – gluttony and carnality – were eschewed by early Christians. Even in modern secularized societies, Lupton (1996) claims, this association persists, while Fisher (1943: 353, quoted in Galef 1994: 6) says that 'It seems to me that our three basic needs, for food and security and love, are so mixed and mingled and entwined that we cannot straightly think of one without the others.'

Meals are often part of dating, celebrating or spicing up a relationship and a prelude to sex – witness Jill Posener's (1987) discussion of food, love and sex in *Turning the Tables: recipes and reflections on women*, a feminist cookbook that combines autobiography with recipes (Bell and Valentine 1995a). Affairs often revolve around eating out. Sally Cline (1990: 61) discusses two cases. In the first, restaurants became an important site for playing out an adulterous affair. As one of the couple commented, '[F]ood assumed a great significance in our lives. It was almost more of a thrill than sex!' In the other case, Gillian, a civil servant, describes how her first lesbian relationship changed the geography of her eating and the intensity of emotion that was articulated through the food she prepared at home for her lover.

> 'I had my first homosexual relationship with somebody who couldn't bear to be known about, so I went underground. There was no way we could eat out together. My social life changed completely. So did my feelings about food. . . . She cooked for me and I cooked for her, quietly, with great love. . . . Those meals certainly weren't routine like the ones with Frank [her ex-partner]. I remember the foods I ate with Edith because of the intensity of feeling.'
>
> Interviewee quoted in Cline 1990: 62

Food not only articulates desire but can also become a substitute for emotional (un)fulfilment. Barbara Pym wrote to Philip Larkin, 'When I was much younger unrequited love caused me to buy and eat half pound slabs of Cadbury's coffee-milk chocolate' (Holt and Pym 1984: 297, quoted in Galef 1994: 4). And so food and sex bring us nicely full circle back to body shape and Lupton's (1996) 'food/health/beauty triplex'; but they also lead on to thoughts about geographies of eating at home.

3

Home

TORSTEN HÄGERSTRAND (32)

•

I wouldn't live up to my role as a 'time-geographer' if I did not maintain that experience of food, like so many things, varies with place, time, mood and company. It is situated, as action theorists say. . . . I like to tell you two short stories to elucidate this matter.

Oatmeal porridge was common for supper in my home when I was a small child. I disliked it utterly, despite the fact that my mother asserted that the stuff settled as 'cotton around the heart'. Sometimes my father had to tell stories and feed me spoon by spoon, one for each sentence.

Years later, when I was 16 and 17, I made a bicycle excursion with two friends on the island of Öland in the Baltic Sea. We studied ancient monuments and looked for rare orchids. One evening at the end of the trip we had to move west for hours. We were blinded by the setting sun and had to battle against a strong wind full of sticking dust. Famished, exhausted and with burning skin we finally reached our hostel at Stora Ror and sat down to eat. Oatmeal porridge was all we could afford. But what a healing and wonderful food the grey mass turned out to be with its thin film of granulated sugar! How could I ever have hated such delicious food? It sucked up the lactic acid from tired legs and indeed felt like cotton around the heart.

Back to childhood again for the next story. When I was about 4 my parents brought me along to some professional conference of which nothing remains in my memory except the concluding dinner party. Organisers were generous enough to let me have a seat at the banquet table beside my mother, I suppose on the condition that I kept quiet.

I must have enjoyed the first courses because when the sweet was in place in front of me I felt too full up to continue eating. I decided to rest a little while. It was a pleasure merely to look at it, a yellow-white pyramidal shape with a shining red cherry on top. I looked forward to attacking the creation with my spoon.

The grown-ups continued to chat and eat, of course. Suddenly it was time for the waiters to remove used plates and prepare for coffee. Unexpectedly I saw my untouched sweet disappear above my head. A modern child would have emitted a loud protest. Such behaviour was out of the question so long ago. So, sad and angry, I lost my wonderful sweet.

That is not the end of the story. The sweet will not disappear from my mind as easily as it disappeared from the table. The odd thing is that it has grown in size over time in proportion to my own body or more than so. The red cherry has got the diameter of a table-tennis ball and still rests at the level of my nose when I think of it at the table. If it were placed in front of me again I would still not be able to eat my sweet. It is too big. The 1920s banquet will forever remain an unfinished symphony.

HOME

What constitutes a 'proper' meal has been a popular theme within sociological work on food. We begin this chapter by focusing on this body of work. This argues that a 'proper meal' is a meal which is eaten together as a family and that part of the intention behind producing such a meal is to produce 'home' and 'family'. We then go on to suggest that a commitment to sharing the preparation and consumption of food can also be important to the production and maintenance of households that are shared voluntarily; and to consider the way that welfare and housing agencies have used food as one way to instruct the homeless poor in a particular way of home life.

The chapter then moves from considering the role food plays in the production of household identities to explore the complex ways in which individual identities are also articulated and contested through food consumption and the spatial dynamics of cooking and eating in the home. In particular, we focus on how patriarchal gender relationships are produced through the preparation and cooking of food, the organisation and structure of meal times, and the design and layout of spaces in the home.

Much of the sociological work on food and families has, however, been subject to criticism for considering the home as a single site of consumption rather than a place for multiple consumption practices; and for focusing on one lifestage experience – that of adults with young children – thus ignoring the ways in which food consumption practices are renegotiated and reproduced throughout the multiple stages which make up an individual's life course. We therefore go on in this chapter to look at negotiation and conflict between adults, and between children and adults, in different household forms. We argue that the home is a site of multiple, sometimes contradictory, consumption practices crossed by complex webs of power relations between household members – and that these in turn both shape and are shaped by the ways in which both individual and household identities are constituted.

Finally, we consider some of the radical visions of new ways of living which characterise many feminist reimaginings of non-sexist housing and home life in which food preparation and consumption take place collectively rather than in family units.

3

HOME

•

The home is one of the most important sites in our everyday lifeworlds. It is where we commonly begin and end the day. It is the place where we carry out all those nitty-gritty bodily chores: sleeping, washing, dressing, and above all cooking and eating. The sharing of these bodily practices, particularly food preparation and consumption, plays an important part in constituting household relationships and identities. This chapter explores the question of 'identity' in relation to the cultural location 'home', focusing on some of the complex ways individual and household identities are produced, articulated and contested through food consumption and the spatial dynamics of cooking and eating.

PROPER MEALS: PROPER 'FAMILIES'

Key distinctions in modern Western societies are commonly made between meals and snacks on the basis of what is consumed, how it is consumed, who is involved, where the consumption takes place and the complexity of the event (Whitehead 1984). A snack, according to Nicod (1974), is an unstructured food event in that there are no rules or protocol about what should be eaten together and in what order. It is composed of one or more self-contained items which are not commonly eaten at routine times and it usually involves limited preparation (i.e. is consumed straight out of the packaging or briefly and unceremoniously relocated from a container to a plate). The consumption of snack foods grew in the UK at a rate of 8 per cent per annum between 1983 and 1988 (Fine and Leopold 1993). This process of nibbling at or grazing on snack foods in place of a 'proper meal' has been seen as individualising eating (Strasser 1982; Fine and Leopold 1993), creating what Fischler (1980: 946) terms 'the empire of snacks'.

A meal, on the other hand, is a physical event. It has no self-contained food items (so everything must be prepared) and is tightly bound by social rules, including those of the meal format – in other words, the sequences and combinations in which specific foods should be consumed (Douglas and Nicod 1974; Nicod 1974). A meal is commonly eaten at routine times and is a social event often involving 'family members'. It is usually prepared in the kitchen and is accompanied by a whole protocol about who should sit where and how they should behave (Whitehead 1984). Thus Douglas (1972, 1984) argues that a meal can both articulate social relations inside a household and define the boundaries between household members and 'outsiders'. For example, who a meal is shared with at home (and when this eating event takes place) serves to distinguish between those who are close friends and those who are only casual acquaintances.

The sociological literature on food is replete with references to 'proper meals'. One of the most widely quoted studies is Charles and Kerr's (1988) research project on food and families, which involved interviews and food diaries with two hundred households in the north-east of England. Their research found that 'a proper meal' is usually defined as food which has been cooked and is hot rather than cold. Merely preparing something, such as salad, or heating something up, like a cook-chilled ready meal, does not, according to Charles and Kerr's respondents, constitute either 'cooking' or a 'proper meal'. A similar definition was also provided by interviewees taking part in Keane and Willetts' anthropological study of healthy eating in south-east London: they argued that a 'proper meal' is 'home-made and requires time and effort to prepare' (Keane and Willetts 1995: 17).

An intrinsic part of these definitions is that a proper meal involves 'good food'. This, according to Charles and Kerr's interviewees, is food which is constructed as 'fresh' and 'natural', the stereotypical 'proper meal' in Britain being 'meat and two veg' epitomised by the 'Sunday roast'. Murcott's (1993b) study of thirty-seven pregnant women in a South Wales valley provides a more nuanced account of 'good food' which also recognises the historical specificity of such constructions. In 'Talking of good food', Murcott (1993b) highlights the plurality of meanings evident in her interviews, focusing in particular on the tension between, on the one hand, women talking about the goodness of food in medical or health terms and, on the other, their discussion of its gastronomic goodness. For many, of course, such as those on state benefit, 'good food' and hence a 'proper meal' have to be adapted to suit economic conditions within the household (Charles and Kerr 1986c). And it has also, of course, adapted to suit changing social and cultural conditions (Hardyment 1995).

Charles and Kerr argue that a 'proper meal' is also a meal eaten together as a family and in fact that a 'proper meal' is 'itself constitutive of the family as a cohesive social unit' (17). They state that 'food is important to the social reproduction of the family in both its nuclear and extended forms and food practices help to maintain and reinforce a coherent ideology of the family throughout the social structure' (Charles and Kerr 1988: 17). Their claim is echoed by other writers working on consumption. DeVault (1991: 79) claims, for example, that 'Part of the intention behind producing the meal is to produce "home" and "family"', while Young and Wilmott claim that the 'unity' of the modern family has been 'restored around its function not of production but of consumption' (quoted in Campbell 1995: 107).

In particular, the dinner table has been identified as an important site for the socialisation or 'civilisation' of children (Plate 3.1). In a historical study, Elias (1978) has traced the way that Europeans, from the sixteenth century onwards, began to regulate their emotions and public behaviour. Gradually, bodily functions – spitting, urinating and so on – which it was perfectly acceptable to perform in public in medieval times became associated with shame and embarrassment. There was also a parallel civilising of appetite. In the eighteenth and nineteenth centuries, self-control and eating in moderation began to be considered important social virtues. These eating practices were contrasted with the vulgarity and gluttony of lower classes (Mennell 1991). Bodily

Plate 3.1 The family that eats together stays together
Photograph: Corbis-Bettmann

management has therefore became important in socially differentiating people and in symbolising the triumph of culture over nature (Lupton 1996).

Thus, as Fischler (1986) points out, the disciplining of children at the table is not just a case of trying to protect children's health by teaching them which foods are edible, it is also about personhood and enculturation: how to prepare and cook foods, what it is appropriate to eat at particular times and how to behave in a civilised manner. Table manners such as how to use the correct cutlery, not to speak when you have a mouth full, not to eat noisily and to keep 'all uncooked joints' (i.e. elbows) off the table are all subtle ways of teaching children to manage their bodies. Widdowson (1981) claims that this civilisation process begins early: repetitive linguistic phrases such as 'pardon me' and 'excuse me' and appropriate sounds and words such as 'yummy' are even used by mothers when they feed babies.

The dinner table also serves to reproduce the 'family' in more positive ways. Dorfman (1992) claims that rather than merely having a traditional culinary function, the modern kitchen has become an all-purpose space, described by one designer as one of the most fascinating rooms of the house. It is at the kitchen table that children do their homework, and we share many of our good and bad times with friends and family there. Dorfman (1992) uses examples from a range of US soap operas, such as *Friends*, *Kate and Allie*, and movies such as *Annie Hall*, *The Big Chill* and *Moonstruck* to highlight the role the kitchen table plays as a site for intimate conversations and the exchange of confidences, as well as sometimes acting as a 'contemporary boudoir' (Dorfman 1992: 36). By acting as a focal point of the day, the kitchen and in particular the shared 'family' meal gives household members a rare chance to come together and catch up on each others' lives. Eating together is a particularly important way of incorporating new household members and fostering a sense of cohesion among reconstituted families (Brannen *et al.* 1994). Meal time is also a time and place where parents have the opportunity to pick up on any problems their children may be having at school or work and to tackle family disputes. Two parents interviewed by Brannen *et al.* (1994) describe the unifying role of the shared family meal thus:

Father: I suppose there is an element of – we're a family and we ought to spend time together and we ought to be able to converse, relate to each other, rely on each other maybe. It's a bit of the old Victorian family.

Mother: I think that's incredibly important. (*Why?*) Exchange of views, I mean she [daughter] wouldn't see the little young men [her brothers] very much at all if she didn't eat with them sometimes. Ummm, I don't know, I think it's quite a good place to discuss things, you learn about each other then.

Interviewees quoted in Brannen *et al.* 1994: 149

Indeed, some of Charles and Kerr's (1988) interviewees argued that families that do not eat a traditional Sunday lunch together are 'not proper families'. This is a particular anxiety for mothers in full-time paid employment. Working parents interviewed by DeVault (1991) felt frustrated and defeated when they could not organise regular 'proper family meals', while mothers in a study by Kirk and Gillespie (1990) felt continually guilty that they were neglecting their children's diet by working outside the home, even though the research showed that this was not the case in practice. One mother, interviewed by Brannen *et al.* (1994), describes her guilt at the fact that her family do not eat together in the week, which she blames on her job:

> 'That's another thing I feel guilty about because they are left to virtually get their own tea ...' (*Do you eat together as a household at all?*) 'Well it's dreadful to say this, but usually only on a Sunday, that's Sunday lunch.... My Mum would turn in her grave, because we were a family that had three meals, we all sat at the table and you ate your meal.'
>
> Interviewee quoted in Brannen *et al.* 1994: 161

A commitment to sharing the preparation and consumption of food can be just as important to the production and maintenance of households that are shared voluntarily as it is to the social reproduction of 'proper families'. Graham (1981) describes the role food played in constituting the 'home' she shared with ten other undergraduate students in west Philadelphia, USA. At initial organisational house meetings explicit ground rules were negotiated and laid down in order to try to make the communal living project a success (Graham 1981). Despite different levels of culinary interest, skill and experience, each member of the household was expected to cook proper meals (pre-prepared foods were frowned on as being less healthy and more expensive) three times per month and wash up three times per month. The evening meal was an important social event which played a key role in constituting the collective identity of the household. Eating together established a strong degree of communion among the inhabitants and even articulated other 'family' characteristics. Special feasts and cakes were prepared to celebrate each resident's birthday and to mark Thanksgiving, Christmas and Passover Seder. However, after several years the membership of the household changed. The newcomers invested less time and effort into preparing and consuming the 'family' meal, and consequently Graham (1981) argues that the spirit of reciprocity, sense of communion and collective identity which had been constituted at the dinner table waned.

The importance of shared meals in the social production of households is further emphasised by the role that food plays in people's memories of 'home'. Morgan (1996) notes that Muriel Spark begins her autobiography with memories of particular foods;

and Richard Hoggart places great emphasis on food in his account of his working-class childhood in Leeds. Morgan argues that

> Food represents a particularly strong form of anchorage in the past, its strength deriving in part from the familial relationships in which the serving and preparing of foods are located.... Food, then, serves as one of the links between historical time, individual time and household time.
>
> Morgan 1996: 166

In a study of childhood memories of food, Lupton (1994) asked Australian postgraduates to write about a memory in the third person. Even though the participants had been set the open-ended task of writing about food rather than meals, many of them actually recalled family dinners. Rather than the content of these meals, it was generally the social relationships that accompanied them which the participants remembered. Several of the accounts described meals as a source of family cohesion, uniting not only the immediate family, but also the extended family of grandparents and aunts and uncles. Some of the participants who had emigrated to Australia from other countries interpreted 'home' on a different geographic scale, using food as a way to articulate their nostalgia for their 'homeland'. Maria, for example, described fish soup and fruits which reminded her of her childhood in Portugal; while red meat and sauces symbolised 'home' for Jurgen, who grew up in Germany (see also Lupton 1996).

Passing on recipes and particular cooking techniques from one generation to another (usually from mother to daughter) is one way in which some households have traditionally reproduced their 'identities' over time, although some commentators have claimed that new food technologies (e.g. the microwave) and convenience foods are destroying this tradition. One group of British Asian women in Ealing, west London, are so concerned that their adult children's diets are becoming Westernised and that they will soon lose the 'taste of home' that they have started a cottage industry in Ealing to provide traditional home-cooked foods for local young British Asians. But the women also realise that this is only a short-term solution because their grown-up children are so busy at college or work that they do not have the time to learn these family recipes and cooking skills from them. Recognising that when they die their cooking skills and recipes will die with them, the Ealing women have now started a community project to put together a book of traditional recipes for their children so that their ethnic identity may continue to be reproduced on the dinner tables of future generations of their families.

CREATING 'IDEAL' HOMES : FOOD AND THE HOMELESS

The importance attributed to food preparation in the reproduction of 'family' and 'home' is also evident in the historical and contemporary response of welfare and housing agencies to the 'homeless'. Veness (1994) describes the US settlement house policy of the late nineteenth and early twentieth centuries:

> [S]ettlement houses were established in immigrant and poor neighbourhoods in order to turn residents of these neighbourhoods into model Americans and proper homemakers. By being good neighbours, by inviting local residents into the settlement house to participate in such programs as food preparation, personal hygiene, and team sports, the settlement house hoped to inculcate those habits and attitudes needed for middle class life later.
>
> Veness 1994: 152

She goes on to outline how this pattern of using welfare and housing policies to instruct the homeless poor of the USA into a particular way of life and regulate their behaviour has continued in different forms to the present day. Contemporary shelters screen potential residents and cream off the more 'acceptable' poor. Women, in particular, are taught about food and hygiene skills in preparation for when they are homed.

While Veness points out the insidious nature of this 'moral' education, it is also true that the homeless tend to have chronic health problems and are at high risk from malnutrition. A study of single women with young children living in temporary housing shelters found that their eating practices were seriously affecting their own health and their children's mental and physical development. High levels of iron deficiency, anaemia, obesity and hypercholesterolaemia were common (Drake 1992). Homeless people who have access to their own cooking facilities have much better diets than those who live in hotels and other forms of temporary accommodation (Wiecha et al. 1993).

At NOMAD, which provides sheltered accommodation for young homeless people, a high priority is placed on ensuring that the residents eat properly. The residents have access to shared kitchen facilities (including basic utensils, a sink, fridge, cooker, storage space and so on) but no food is provided for them, except when they first arrive. A key worker (see Box 3.1) helps all new arrivals with budgeting and takes them out on shopping trips, advising them what they should buy. Most meals are prepared individually but the staff organise a session where all the residents cook a meal together. The aim of these sessions is to ensure that all the residents eat properly, but it is also a subtle way of teaching them food hygiene and cooking skills as many of the young people who come to the centre have had only very unstable and fragmented experiences of 'home', and consequently have little experience of eating 'proper meals' or of eating with

Box 3.1 KATHERINE

KATHERINE IS A WORKER AT NOMAD. SHE DESCRIBES THE ROLE THE STAFF OF THE RESIDENTIAL PROJECT PLAY IN ENSURING THAT THE YOUNG HOMELESS RESIDENTS EAT PROPERLY.

'Well, I mean, the residents are all from different backgrounds, I mean, family breakdowns, um, it depends how long they've been on the streets, some of 'em, some of them could have been on the streets since [they were] 10 so would never sort of, you know, like for me to have me mum and dad to cook a meal, they've probably never know that, so they've always had to cook themselves, so that's probably why some residents just like to eat out of tins and grab what they can get. Plus, um, shopping for them isn't a number one priority. So some shop and some don't, you see. The same with the hygiene, you know, that's at the bottom of their list as well so that's, well, our job, turn it around.

'We help them along with food so then we know that they're eating, and then we have a cooking session and staff bring food in and we buy food out of petty cash and just like cook a meal and that together.

'I mean, er, one of the residents we've got in at the moment eats a lot of chips so we sort of, it's like laying, I know it's probably awful but it's like laying traps down and, er, you turn it round and you ask them what do they like to eat and they might say, oh um, I like so and so, "Well you know, why don't you buy so and so then?", 'cos they can't be bothered to cook it, "Well buy it and we'll help you cook it, you know, together", we sort of get round things that way.

'I mean, if we know that they're not eating or they are eating but they're not eating sommat substantial then we'll come in and we'll make a meal and then we'll all have it together. I mean a lot of people, well not a lot but say some, don't like eating in public, you know they like to eat on their own, or a lot of them could put food down as their way of getting attention out of people, you know, they could turn round and say, "I've just been sick", for attention off workers and other residents, or they could say, "I don't like that food" but really they love it, but they just won't eat it because there's other people watching.'

and in front of other people. For some of the residents their conceptions of an ideal home and their imaginings of what it would be like to have their own family are very much tied up with an imagining of sharing a roast Sunday lunch, as Sharon describes in Box 3.2. In this way food articulates several meanings of home, including what Sommerville (1992) termed home as a haven – a place of physical security; home as hearth – a place with a homely atmosphere that welcomes others; and home as heart – a place of emotional stability based on relations of mutual love and happiness.

Box 3.2 SHARON

SHARON, A NOMAD RESIDENT, DESCRIBES THE SORT OF FAMILY MEALS SHE
PLANS TO HAVE WHEN SHE HAS HER OWN HOME.

'This isn't my home, you know, I'm cleaning somebody else's home, and it's like if I had
me own I'd be – oh, spotless! But, like, I just think it's a waste of time doing it in everybody
else's house 'cos it's just, it's not mine, you know what I mean? And I don't bother, it's like,
that's why we don't cook much up there 'cos it's not yours, you know, it's someone else's
and it's not the same, but, like, you just think, "oh well, I live here, don't I, somebody'll
have to do it."'

'When I moved in with me little sister they had their family had Sunday dinner all together,
it's like the son'd come up and, like, their Sue and the kids would probably stay at home,
you know, and have it as their own family. And I felt a bit uncomfortable at first, you know,
eating in front of everybody but we'd all have a big Sunday dinner together, and we used
to have a right laugh, 'cos we used to go to the pub after, you know, like go to the pub and
come home, have Sunday dinner and then go back to the pub for half an hour and get us
head down, you know. Sunday dinner, get your head down. But we used to have some right
laughs, you know, 'cos like everybody were talking about different things, and it were right
good. That's why I want a big, I want a big family, me, 'cos I prefer, you know, I like fussing,
like I'm one of them 'cos like my family never did it with me, you know what I mean? And
I like to be, I like to fuss people, you know, to show me appreciation.'

THE WAY TO A MAN'S HEART IS THROUGH HIS STOMACH!

So we have to attend to the social relationship within which eating, cooking, the
expression of choice are located and appreciate the social organisation of these
activities and relationships. So we have to look out for questions of autonomy,
control, power, the exertions of the sanctions in such relationships, and
questions of ideology (in its minimal sense) and belief that overlie these more
fundamental dimensions.

Murcott 1986: 87

In pre-industrial Europe production (baking, weaving, farming, etc.) and reproduction
(cooking, eating, sleeping, child rearing) took place in the same location. There was no
separation of activities (work and home) into different spaces (public and private).
Rather, people lived communal lifestyles with women and children playing an active part
in economic society. Following the development of industrial capitalism in the nineteenth
century, however, reproduction was removed from the communal sphere and relegated

to the private sphere of the home. Activities such as preparing food, cooking and eating became more privatised, and new ideologies about gender roles emerged (England 1991; McDowell 1983). It was women who were seen as having all the necessary attributes to manage the home and provide emotional and bodily care for other members of the family.

Despite the changing nature of gender roles and gender relations in the twentieth century it is well documented by sociologists in the 1980s (Luxton 1980; Edgell 1980; Collins 1985; Charles and Kerr 1986c, 1988; DeVault 1991), particularly Anne Murcott (1982b, 1983a, 1983b), that within families it is usually women who still shoulder the responsibility of choosing, shopping for, preparing and cooking 'proper meals' for the 'family' (Plate 3.2). Men are still regarded as the providers, women the nourishers (Blaxter and Paterson 1983).

Studies of households in South Wales by Murcott (1983b) and of families in York by Charles and Kerr (1988) found that men never took responsibility for the day-to-day cooking. Their involvement in the kitchen was strictly limited to casual, simple tasks. This pattern is largely corroborated by a survey conducted in 1995 by *Sainsbury's Magazine* (a UK supermarket magazine). This found that, in the majority of the 43,000 households that returned the questionnaire, women (even those in full-time employment) were primarily responsible for shopping (62 per cent) and cooking (75 per cent). A slightly different picture is presented, however, by a recent academic study set in Greater Manchester (Warde and Hetherington 1994). This found that of those men questioned whose female partners were in full-time employment (a household pattern that comprised a small sample in Charles and Kerr's study), 25 per cent of them had prepared the last household meal. A journalist writing in *The New York Times* Living Section also claims that men are taking on more responsibility in the home. He suggests that contemporary men not only cook, but even trade recipes:

> It's a notion as outdated as the idea that men don't cook. Of course, men swap recipes as well, at the gym or bars, over fax machines or card games, during half-times or seventh-inning stretches – or even over clothes lines.
>
> Alexander 1991 quoted in Dorfman 1992: 34

The majority of academic research, however, continues to suggest that the level of men's participation in the preparation of food does not correspond with the popular impression that patterns of the gender division of labour are changing. Rather, domestic divisions of labour in the kitchen remain much as they did thirty years ago, with men pulling their weight only when they have to (e.g. because their partner is at work) (England and Farkas 1986; Warde and Hetherington 1994).

The traditional division of labour appears to be particularly evident among

Plate 3.2 Women shoulder the burden of feeding 'the family'
Photograph: Corbis-Bettmann

'working-class' households, where men and women appear to have more rigidly segregated tasks; middle-class households appear to aspire to a more shared lifestyle (Young and Wilmott 1975; Oakely 1974). Indeed, on the basis of limited research, middle-class men do actually appear to carry out slightly more domestic tasks than men in lower-income households (Charles and Kerr 1988; Brannen and Moss 1991). Warde and Hetherington (1994) found that women in what they termed 'pure' working-class households were responsible for the majority of the cooking, while those in 'pure' middle-class households did the least work in the kitchen.

Brannen and Moss (1991) argue that men's low participation in domestic chores is reflected in women's low expectations – with both sexes persisting in the view that men's paid employment is more important than women's. Charles and Kerr's (1986c) respondents argued that men deserve a 'proper meal' after a hard day's work, and often excused men from cooking on the grounds that they are so incompetent at domestic chores that it is not worth making them cook (Mansfield and Collard 1988). In contrast, in Sweden, where there is a higher degree of gender equality in paid employment (in public), there is also a parallel expectation of greater gender equality in the home (in private). In a study by Jansson (1995), men in half the households surveyed actually claimed to cook a 'proper meal' at least once a week and to take their turn food shopping.

When men in Britain do help in the kitchen it is generally regarded as a bonus or a treat (Charles and Kerr 1988). On these occasions men usually cook or prepare the things that are stereotypically enjoyed by men such as coffee, alcohol, pies, meat and potatoes (Adler 1981). Meat is a particularly masculine food while vegetables and salads are perceived to be feminine (Shapiro 1986). MacClancy says:

Meat is strength. Meat is power. Meat is life. It is the very king of foods. It gives us might, increases our potency, adds edge to our aggression, heats our passion, augments our sexuality, and turns us males into macho men.

MacClancy 1992: 145

The kitchen itself has taken on a feminine gender identity (Lowe, Foxcroft and Sibley 1993; Putnam and Newton 1990). In contrast to more ambiguous spaces within the house which may have multiple uses and may be dominated by different members of the household at different times of the day, such as the living-room, the kitchen in many 'family' homes is strongly defined as a woman's space. Feminist architects have traced the evolution of the design of contemporary housing, demonstrating that rooms associated with men are always focused at the front of the house – the high-status, 'public' section of the home – while rooms which have historically been associated with women, such as the kitchen, are hidden away at the rear of the house (Matrix 1984; Roberts 1991). Thus they argue that this front/back, public/private dichotomy articulates patriarchal gender relations. Different cooking techniques are also highly gendered. Boiling is perceived to be feminine, the techniques of roasting or barbecuing masculine. Men's everyday participation in feeding the family is therefore usually limited to a few carefully chosen and bounded activities: running the barbecue, carving the roast, making alcoholic drinks, fetching the take-away (Adler 1981; Warde and Hetherington 1994) and, in the USA, cooking pancakes for Sunday breakfast (Adler 1981). As one writer explains:

> [T]he kitchen had come to be regarded as woman's sphere from frontier days onward, [but] cooking out of doors was different. It reminded grown men of Boy Scout days when they roasted hot dogs over campfires. Building up a good fire in the charcoal grill was thoroughly masculine. So was the cooking of a steak.
>
> Watson 1962: 318, cited in Adler 1981: 46

Men, Adler (1981) claims, love to turn their rare forays into the kitchen or the barbecue pit into a hobby, buying special cooking gadgets, making a fuss about their special recipes, emphasising all the complexities of what they are cooking and turning its consumption into a special event rather than a routine meal. As a result, some men's efforts in the kitchen can actually cause more domestic tension than their non-participation, with some women defending the boundaries of the kitchen as their own space against the encroachment of male 'hobby' cooks. As a woman interviewed by Cline explains:

> 'When my husband deigned to do a curry I spent the entire bloody Saturday in the kitchen chopping and preparing and the entire bloody Sunday clearing up. He spent one hour stirring, tasting and prima donna-ing and everybody said not only "Oh what a wonderful meal" but they actually turned round to me and said "Gee you're lucky to have a husband that cooks"'.
>
> Interviewee quoted in Cline 1990: 72

Domestic routines usually centre around men; as Charles and Kerr (1986c: 64) comment, 'Men, although they do not stir the cooking pot, control to a significant extent what goes into it.' Similarly, Murcott (1983b) suggests that whether or not a woman is going to go to the trouble of cooking a 'proper meal' usually depends whether and when her husband is going to be home. Women will not go to a lot of trouble if the meal is just for the children and themselves. Women also usually privilege their partners' and children's tastes over their own (DeVault 1991), especially in low-income and lone-parent families because it is seen both as a way to avoid wasting food and money and as a way of caring for the family (Graham 1987). Learning what men like by trial and error is therefore, for women, an important part of setting up home together (Murcott 1983b; Charles and Kerr 1986c, 1988).

The case study in Box 3.3 provides a good example of the way men's needs dominate the organisation of cooking and eating. When Paul was a child his meals were structured around his father's and elder brothers' shifts at the local pit. Shortly after he left school

Box 3.3 PAUL

PAUL WORKED AS A MINER UNTIL HE SUSTAINED SPINAL INJURIES IN AN ACCIDENT. HE IS NOW UNEMPLOYED, AND SPENDS MUCH OF HIS TIME TRAINING WITH A WHEELCHAIR BASKETBALL TEAM. HE LIVES AT HOME WITH HIS WIDOWED MOTHER.

'I used to eat with my parents sort of thing because me brothers were working, they used to work down mines at pit so we didn't used to eat very often as a family 'cos they were on shift work, you see, so they were never at home at same time. But then when me Dad used to work at, he used to do shifts . . . it depended really what shift they were on and when they were there and plus if I were doing anything after school, whether I come late or stop back for sports or whatever. So – used to eat at weekends as a family but in week it used to be, used to vary because, er, shift work, like I say.

'I ate a lot then – when I first went home after accident, 'cos I put a lot of weight on, but that might have been because I weren't doing a lot as well, 'cos I weren't doing anything or nothing really, no exercise or sports really for a bit, so – er you just used to eat, say like then there were only me at home you see as well, me other brother had, er, gone, been married then. So I'd have ate when me Dad had ate, sort of thing, I think, you know when he come home from work, I'd have waited for him, I wouldn't have said I wanted mine at four o'clock if he were coming at six, I'd have waited till six.

'When I had me accident it, I had my accident in June – me Dad had a stroke in Christmas straight after that, you see, and they say it might have been a bit connected with it, you see – and, er, he were never right, so he didn't work then either, you see, from that time onwards. . . . I think we really, we'd have ate as a family more then, than we ever, ever did really, 'cos we were all in house at same time sort of thing . . . and then it had changed when I were basketballing – 'cos they'd have ate on their own – and I'd have ate separate sort of thing when I'd come in or before I went out or if I were in I'd have ate with 'em.

'Well I'm only, only me Mum now you see, so me Dad's died, so, er – I'd have said all meals at home now go round, go round me – like, it's gone, it's gone to me sort of thing, so really through time it's gone through different people. . . . Thinking about it . . . it goes round me . . .

'Sometimes we eat different meals, we'll still eat at same time but we'll eat different meals. . . . I cook meself, I'll cook meself pasta and stuff like that and I cook, cook for meself a fair bit now really; if I'm doing anything like that, say, she says, "I'll do that", and then, but she'll not eat, she'll not do herself a cooked meal, she might just have a sandwich or something like . . . we always still have a Sunday lunch . . .

'Yeah, I sit at, we sit at table 'cos I, I'd sooner eat at table in me chair sort of thing now. Like I said, before me accident I used to be in room all time, but now I'd sooner sit at table. Er – that's might be 'cos I'm in me chair, it's a bit awkward to have a tray on your knee while you're eating in a chair, so – and she always sits at table with me, so – we do sit at table. If there's something on telly, sportwise, more or less I'll put telly on while we're eating 'cos I like sport so I watch a bit of sport. Er – and that's it really – we do sit, sit at table.'

he also started to work at the same colliery but then suffered a serious spinal injury in an accident which confined him to a wheelchair. Unemployed and at home with his mother, Paul and his family continued to structure their meals around his father's work and later his father's illness. After his father's death Paul began to take up wheelchair basketball, competing at a high level. Now meals in the household are oriented around his training routines. Paul is very conscious of the relationship between sporting performance and diet, and eats a lot of pasta, which his mother often cooks for him. She does not like pasta and so when she prepares it for him she just makes a sandwich for herself. Despite all the changes in the composition of the household and its eating habits, Paul's mother has continued to organise her meals around her menfolk.

Women's role as the 'cook' is a symbolic as well as a material task. It has been argued that women can derive both pleasure and identity from providing 'cooked dinners' and meals which suit the tastes of individual household members (Murcott 1983b). For example, Charles and Kerr's (1986c, 1988) interviewees said that while men should be able to cook, they should not have to, because it is 'women's work' and cooking a proper meal for their partners was a way of articulating their love for him. The way to a man's heart, it appears, is still through his stomach. In a detailed and nuanced paper, Dorfman (1992) has traced the history of this message, demonstrating how food industries, advertisers, women's magazines and cookbooks have all played a part in reproducing the idea that food is important in creating and maintaining a successful marriage and home. The flip side of this, of course, is that women's failure to provide 'good' cooking is often used by men to justify domestic violence against them (Dobash and Dobash 1980; Ellis 1982), as this woman remembers from her childhood:

'Childhood was a series of frightening meals. Mother was always humiliated. Father shouted about the food, then my sister wouldn't eat, and Mother thought she'd failed. Because she felt pressurised by Dad who despised her cooking, she insisted we ate everything up. There was always a bad atmosphere at meals. You never knew how much he'd shout about tea being on time, or if it would get worse. Meals were dangerous places.'

Interviewee quoted in Cline 1990: 147

Sociologists have therefore argued that the meal in domestic life can articulate not only the identity of the 'family' and 'the home' but also gender roles, identities and power relations between different members of the household.

RENEGOTIATING HOUSEHOLD IDENTITIES

Recently, the literature on 'feeding the family' has been subject to a growing number of critiques. Campbell (1995: 107), for example, points out that this body of work often implies that the family or 'household is regarded as if it constituted a single unit of consumption'. He thus argues that this literature 'ignores complex intra-familial processes which in practice directly affect consumption [and that] it is necessary to understand decision-making processes within the family if one is fully to understand how the home functions as a consumption "site"' (Campbell 1995: 107).

Delphy (1979) has suggested that we should think of the home less as functioning as a single unit of consumption and more as a distribution centre for its household members. She argues that different forms of consumption take place in the home: collective consumption where everyone is present and eats the same meal; individual consumption which occurs regardless of whether anyone else is there (for example, eating a sandwich alone in the lounge); and individual consumption that is not shared with other members of the household even though they are present in the home.

The focus on traditional nuclear families in much sociological work on domestic consumption has also come under fire. Stacey argues that the family is 'an ideological concept that imposes mythical homogeneity on the diverse means by which people organise their intimate relationships [such that] "the family" distorts and devalues this rich variety of kinship stories' (Stacey 1990: 269, quoted in Katz and Monk 1993: 29). She suggests using the term 'postmodern families' to capture the diverse, fluid and complex composition of modern households.

The majority of sociological work on food also focuses largely on one lifestage experience – that of adults with young children – and thus has ignored the way household and gender identities, and food consumption practices, are renegotiated and reproduced throughout the multiple stages which make up an individual's life course (Beardsworth and Keil 1990; Prout 1991). Indeed, it is argued that we spend a relatively short span of our lives in 'nuclear families' and that little attention has been paid to childhood and adolescence, the lives of older people, single-person households, and the processes of transformation which take place as people establish and terminate relationships with others. In particular, sociologists have pointed out the tendency within academia (and indeed everyday life) to treat children as less important than adults and to ignore children's understandings and experiences of their lifeworlds, particularly their sophistication at managing their own space, time and social relations (Waksler 1986; Alanen 1990; James and Prout 1990).

Sensitive to these shortfalls, research by Goode, Curtis and Theophano (1984) into Italian-American familial food habits focused on a number of significant household differences affecting the patterning of meal formats: presence or absence of husband, stage in family cycle, school and work schedules, social networks and the generational cohort of the senior woman in the family. To these they added family-specific preferences and habits, arriving at a finely textured account of one community's many menus (all of which continued to articulate for the community's members an idea of 'Italianness'). Sharman (1991) has also argued that in order to unpick the important changes which occur within a household over time it is crucial to look at detailed life histories. As she writes, 'the particular dietary practices of domestic units at given times can only be fully understood by looking at the processes by which these patterns developed, as well as the current circumstances with which they are associated' (Sharman 1991: 193). Similarly, a small study by Warde and Hetherington (1994: 765) explicitly addressed the critiques of work like Charles and Kerr's (1988, 1986c) for underplaying the life course, concluding that 'life-course stage is highly significant' in terms of who took responsibility for domestic tasks around food.

Elsewhere we have used food biographies to demonstrate the complex role the preparation and consumption of food at home plays in the production of individual and household identities throughout the life course (Bell and Valentine 1996). Our research shows that changes in identity (e.g. from child to independent student; from accountant to environmental worker; and from sharing a house to living alone) are articulated on individuals' plates – affecting not only what is bought to eat and the places from where it is purchased, but also who has prepared it and the spatial dynamics of when and where it is consumed within the home. We have also used personal food histories to show how individuals' eating habits have in turn shaped their identities. For example, one woman's drift into vegetarianism (as a result of her changing access to places to shop and changes in her income) shaped her politics, her employment, her social networks and the constitution of her home. This detail counters the narrow focus of work such as Charles and Kerr's (1988), which assumed a fairly uniform life path for women of marriage and motherhood.

There has also been growing speculation about the extent to which sociological understandings of food and families may be shifting as a result of changing patterns in food production and consumption (Goodman and Redclift 1991; Fine and Leopold 1993) and a number of changes in contemporary Western lifestyles, particularly a trend towards individualised patterns of eating. Vegetarianism is one example of an eating practice which is commonly a personal lifestyle choice rather than a family matter, requiring different meals to be provided for different household members (Warde and Hetherington 1994). Conversion to vegetarianism can cause bewilderment and hostility among family and friends. While some households are tolerant of change and are prepared to accommodate one member's lifestyle change, others put pressure on the convert to conform to the family's foodways. Eating separately can be perceived as not only a rejection of the household's food, but by implication also a rejection of 'the family'. These tensions are particularly apparent at celebrations, such as Christmas, where by tradition extended families come together and unite around the consumption of special meals (Beardsworth and Keil 1992). Kim is a strict vegetarian who eats mainly organic food but lives with Bob, whose favourite food is a rare steak. She describes the problem of sharing a home with someone who has such a different lifestyle and diet:

'[W]e have argued because of our different attitudes. In our first two years, he used to deny he'd eaten a beefburger, when I could smell that he had. I'd refuse to kiss him. Our biggest argument, which we've had at least 30 times, is about how we'll bring up our kids when we have them. He says his children are going to eat meat and have a choice and I say the baby I carry will be the same as me.'

Kim, quoted in Rooke 1995: 30

Serious participation in sport, enjoying a healthy lifestyle, having ill-health, food allergies or just different tastes are some of the many other reasons why people who live together do not necessarily eat together. For example, Diane prefers healthy foods while her partner Timothy is a junk food addict. He explains:

'Diane keeps trying to make me eat salads but salad doesn't taste of anything and I'm hungry half an hour later. I've always eaten junk food. . . . When we go to the supermarket I'll buy boxes of 20 beefburgers and put them in one of our three freezers, which are filled with my food – garlic bread, potato waffles, bubble and squeak, steaks, burgers, chocolate fudge cheesecake, apple pie, tutti-frutti ice cream, chips, breaded mushrooms with garlic. In the evening we eat in shifts. Diane cooks, then I eat in front of the TV, while she cooks something for herself . . . I'd like to sit down with the same meal but she won't.'

Timothy, quoted in Rooke 1995: 32

Our own research suggests that the life patterns of postmodern families are also increasingly complex and diverse, with the consequence that even if they would like to, it is becoming less easy for household members to co-ordinate their schedules so that they can sit down and eat together. The two case studies, Boxes 3.4 and 3.5, illustrate some of the problems of trying to integrate different patterns of work into home routines. Margaret and John established separate eating schedules when they first got married because of their different jobs. She was a teacher and came home hungry, wanting to eat early at 6 pm; he worked at the university, often staying late into the night to use a computer. This pattern of eating apart has persisted even though they have now retired: rather than work late, John now goes to the pub instead. Jon and Claire's jobs also stop them sharing their weekday meals together. They work in different cities and so cannot

Box 3.4 MARGARET AND JOHN

MARGARET AND JOHN ARE MARRIED AND HAVE TWO ADULT SONS. THEY ARE BOTH RETIRED. THE DIVERSE EATING PATTERNS THEY DEVELOPED AS NEWLYWEDS HAVE PERSISTED INTO THEIR RETIREMENT.

Margaret: *And then in the evening – here we differ!*

John: Here we divide.

Margaret: *Because I cook a meal in the evening, and I usually eat at about, I should think, six thirty, seven. My husband is usually out during that period at the local ...*

John: Evening bar.

Margaret: *Hostelry, that's right. And he comes in about an hour later and I usually microwave the meal for him. The same thing warmed up really.*

John: It goes back a long way, because my wife was a teacher, and you used to get home pretty well exhausted, didn't you, about four o'clock, quarter past four. And in those days I was a research student, research fellow, and I used to work really quite long hours, and get home very late.

Margaret: *He used to work through the night sometimes.*

John: But then there was no evening bar involved, I used to come home really quite late. So from the word go, we've never eaten together, have we? At the weekends we do, not during the week. But now it's not a question of me working.

Margaret: *Oh we do vary it a bit. Sometimes I wait for you to come in, don't I? Very occasionally.*

Margaret: *I used to pick the children up from school and then we'd all have our dinner together ... they'd be home for about quarter past twelve, back for about half three to four, and we'd eat about quarter past five. So it all worked out all right actually. I used to start from scratch when John came in ... if I'd had a microwave I wouldn't have done that! I do things that can go in the microwave now. We still don't eat together very often in the evenings ... I don't know whether we're usual or unusual really. I always ate with the children. I don't know whether I should have done that, perhaps not. I don't know. My daughter-in-law doesn't do that. She feeds the child separately. And then she'll eat when her husband comes in later at night, you see. But I don't know. I think children ought to get used to sitting round a table with you. That's my idea, but there you are. Otherwise if you feed a child on its own, it doesn't get the idea of food being part of a group, does it? It's a social occasion, isn't it, really?*

Box 3.4 Continued

John: You see, with Margaret having been a teacher she'd been used to eating fairly early anyway, let's say half past four, that sort of time, so she continued to eat at about half past four with the two lads, which would have been too early for me anyway. But it was made worse, as I say, by working really quite long and unsocial hours on the computer. In those days we only had one computer, housed in the United Steel buildings and we were allowed to use it at night....
I got into a funny pattern of work. Go into work about lunchtime and having been working throughout the night and barely getting any sleep, but of course I'd be back at the United Steel Company maybe six o'clock, seven o'clock that night. I didn't see the kids, really.

live together full-time, only sharing a home and their meals at weekends.

New technologies in the production, processing, storage, packing, marketing and distribution of foods have generated a constant supply of a wide variety of ready-made convenience foods which are available at all times of the day and night (Jerome 1981). These changes in the foodscape are shifting the domestic work of food preparation and consumption from the home to the formal market sector of the economy (restaurants, cafes, take-away fast foods and so on). Warde and Hetherington (1994) suggest that as a result there is some evidence that people are consuming fewer shared meals at home and more what they term 'street foods'. Their survey revealed that 34 per cent of the households questioned buy take-away meals at least once a week and that 12 per cent of those surveyed dine out at restaurants at least once a week, with 9 per cent eating pub meals on a weekly basis.

This growth of the foodscape, combined with advances in domestic technologies (such as the microwave) which are making food preparation in the home quicker, cleaner and simpler, mean that it is easier for contemporary individuals to prepare their own meals to suit their own schedules and lifestyles regardless of their culinary skills (Mintz 1984). According to Jerome (1981), the frozen meal is the 'modern staple emergency meal' for US families (frozen meals are rarely served on special occasions or to guests). It follows from this that meals may be less significant as what Warde and Hetherington (1994: 773) term 'a collective time-discipline' in the way that coming together to eat as a household once structured the day of its members; and that these new technologies may also be leading to the renegotiation of power relationships within the home.

The microwave has been particularly discussed for its role in liberating teenagers from the constraints of the 'family' meal, giving them more options to eat independently from adult control (Bryant et al. 1985; Bull 1988; Prattala 1989). This has prompted

Box 3.5 JON

JON IS 25, AND LIVES IN A SHARED HOUSE IN SHEFFIELD. HIS PARTNER, CLAIRE, LIVES IN LEICESTER AND THEY CAN SPEND TIME TOGETHER ONLY AT WEEKENDS.

'I suppose it is different, shopping with Claire, in a way, because she prefers to have a trolley and I prefer to have a basket! Well, I actually enjoy pushing the trolley around but I feel a bit of a prat pushing a trolley around on my own and picking my own things and putting it into this trolley ... somehow it doesn't seem like one person should have access to a whole trolley! I don't know, I just feel like I'm taking up too much room in the supermarket whereas if there's two people you're more justified in having a trolley.... I mean, unless I know that I'm purposely going to go in and buy three or four packs of lager or something, or something that's going to be heavy – which isn't particularly often – but unless I'm going to get something I know is heavy then I won't get a trolley, whereas you know Claire always will, which is weird ... so I mean sometimes I do end up walking round with, you know, a basket in one hand and five pounds of potatoes in the other and a packet of crisps sort of shoved under my chin or something, or sometimes I go in and not even take a basket and end up staggering around with, sort of, armful of things 'cos I ended up buying more than I thought ... and I suppose shopping with her I tend to, tend to be a bit more extravagant perhaps as well rather than if I'm shopping just for myself. I think this'll make do, I'll manage with just this.

'... generally in the week when I get home I'll eat on my own, make my own food.... I can never be bothered to cook as elaborately if I'm just cooking for myself so I suppose I mind eating on my own in that respect and I don't have as good a meal as I might have, um, but no, I mean, I don't, I don't actively prefer to eat with somebody or on my own, I don't think, apart from just the quality of the food I might have.

'At the weekend I'll either eat with Claire or Claire and some other friends. Yeah, I mean, I rarely eat on my own at weekends, I suppose.... I'm more likely to be involved in doing things other than just getting home and thinking "what do I fancy to eat" 'cos maybe we're going out or maybe some other friends around, and then maybe somebody'll suggest having a take-away or going and getting something from a supermarket or somebody'll offer to cook and so, yeah, it does become more of a social thing rather than just an "I'm hungry so let's eat something nice" type of thing.... I think there is a bit of a sudden contrast and it does seem a bit of a noticeable change that suddenly you're eating with somebody else, and I tend to make a bit more effort than I perhaps would do otherwise to make conversation, 'cos I'm used to just shovelling food down on my own. I think eating together, it's almost the cooking together that's the bit that's so different and we make something more elaborate and sort of have a bit of a banter as we're cooking in the kitchen.'

some popular concern that young people are becoming a generation of nutritionally deficient snackers (Bull 1988; Greenwood and Richardson 1979). A study of young people, health and family life by Brannen *et al.* (1994) found that one-third of the parents and children surveyed rarely eat together and that three-quarters of the young people questioned cook or buy their own meals on a regular basis. Brannen *et al.* describe a range of different households where children for different reasons take responsibility for feeding themselves, often eating irregularly, skipping meals and consuming diets high in fats and sugars and low in fibre. Studies of the elderly have also found that people who live alone and eat alone have poor appetites and poor diets; like teenagers, they eat large quantities of convenience foods at irregular times (McIntosh, Shifflett and Picou 1989).

Other domestic technologies (such as televisions and computers) also allow or encourage the dispersal of postmodern family members to different rooms, or enable them to pursue different meal activities in the same space. A survey by *Sainsbury's Magazine* (1995) found that 1 in 4 of the 43,000 readers responding to the questionnaire almost always eat their evening meal in front of the television. TV programmes can play an important part in structuring the timing and spatial dynamics of eating. In some households meal routines are carefully co-ordinated around the TV schedules so that everyone can eat together at the table; in other households the television is divisive, causing disputes about where and when meals are cooked and eaten. Thus, rather than bringing people together to facilitate family cohesion and unity within the home, domestic technologies may be encouraging diversity within and dispersal of the household (Livingstone 1992). This in turn may be facilitating the renegotiation of gender identities and household identities, and the (re)construction of what is understood to constitute a 'proper meal' (Cockburn and Fürst-Dilić 1994; Jackson and Moores 1995).

In particular, it is important to understand how patterns of eating are negotiated and contested within households in order to understand how the home functions as a consumption 'site'. The participants who took part in Lupton's (1994) project on childhood food memories recalled that meal times were often a power struggle with parents over bodily habits, especially the consumption of 'good' and 'bad' food. Some of them particularly remembered being forced to eat food they did not like, describing the revulsion, resentment, anger and powerlessness they experienced on these occasions. Paul, for example, described his father as a 'vegetable fascist', recalling how he was not allowed to leave the table until he had eaten his peas. Others remembered the strategies they employed to defy parental authority. Gary, who also had a pea phobia, used to smuggle his peas out of the dining room when his mother's back was turned.

Television advertisements for convenience foods and marketing by the food industry encourage children to resist parental definitions of 'good' food and to attempt to redefine the family's diet by nagging their parents for fashionable foods. Food manufacturers are known to put pressure on schools to adopt materials which they produce for teaching. One of the newest vehicles for commercially sponsored school aids is the schools Internet. Nestlé, for example, sponsors 'Insights into the Developing World'. Not content with these ways of reaching potential young customers, some multinationals have even been lobbying the British government to be allowed to pay schools to advertise their products directly in classrooms. As Fischler argues, 'Parental authority is being increasingly subject to competition from such extra-family influences such as school, the media and commercials, food habits of younger generations are no longer shaped by coherent, traditional matrilocal culinary patterns' (Fischler 1980: 949).

Children use a range of negotiating strategies to get their own way, including begging, bribery, bargaining and ultimately tantrums (Walker *et al.* 1994). Parents do not usually like to admit to giving into their children's demands but often have some strategies for compromise up their sleeves. In low-income families it is often a case of parents not being able to provide the expensive food brands that children want. Middleton and Thomas (1994) give the example of a mother filling an empty Frosties packet with a cheaper cereal brand to fool her children into thinking they were eating what they wanted when they were actually eating what she could afford.

Chapman and Maclean's (1993) study of Canadian teenagers found that 'healthy ' foods were associated with being in the parental home. Eating junk food was therefore one way in which young people could articulate their independence from their parents. Food is also an important medium for the expression of hybrid identities. Brannen *et al.* (1994) describe the experiences of two young British Asian women, Aruni and Nasreen. Aruni chooses to prepare most of her own meals in the week, rather than eat with the other members of her household. She primarily prepares convenience and frozen foods for herself instead of Asian dishes like those her mother makes for the rest of the family. For Aruni, eating alone is an opportunity for independence from her family, which she describes as restrictive and over-protective. Nasreen also avoids eating with her family. Like Aruni she describes it as oppressive and restrictive – particularly because she has to behave as a perfect Muslim although she does not believe in God. However, her parents make her eat with them at weekends because they are concerned that she may be on the verge of anorexia. Her father explains:

'I try to eat the evening meal with the household. My daughter doesn't comply. It's part of her attitude to food, her attitude to freedom. Generally she likes to please herself, eat what she likes. I've tried to persuade her eating isn't just about eating, it's a social situation . . . Nasreen doesn't comply. It's part of her attitude to food, her attitude to freedom. . . . In the weekend I apply a little more pressure.'

Nasreen's father, quoted in Brannen *et al.* 1994: 152

Elsewhere we have demonstrated not only the importance of children's agency in household decision-making processes and children's ability to resist adult constructions of 'good food' and 'proper meals', but, more importantly, that in some cases children actually have the power to (re)negotiate the consumption practices and identity of the entire household, persuading, for example, the whole family to become vegetarian (Bell and Valentine 1996).

Academic discussions about conflicts between adults over cooking and eating usually focus on gender relations (as discussed earlier in this chapter), but food practices and consequently household identities are also the subject of negotiation and conflict between adults living together in other forms of relationships (e.g. as friends or as lesbian/gay partners). In Box 3.6, Kate explains how her eating habits have changed each time she has entered a new lesbian relationship. Sometimes these transformations have been smooth, at other times they have been discordant. Sometimes she has imposed her own food habits on to her partner, at other times she has accommodated her partner's tastes or needs. Other research has similarly shown how food comparisons, both favourable and unfavourable, about what constitutes a 'proper meal', the right way to cook particular foodstuffs and the spatial dynamics of eating play an important part in the negotiation of new household identities in reconstituted families (Hughes 1991).

All the examples outlined in this section of the chapter demonstrate that the home is a site of multiple, sometimes contradictory, consumption practices, crossed by complex webs of power relations between household members (not only between men and women, but also between adults and children, same-sex partners and so on), and that these in turn both shape and are shaped by household consumption practices and the ways in which both individual and household identities are constituted. In the light of this material, Anne Murcott's assertion that meal times merely 'dramatise pre-existing relationships' rather than shaping or creating new relationships (quoted in Hardyment 1995: 199) clearly needs refining.

ALTERNATIVE PRACTICES OF 'HOME'

In the 1970s some radical feminists began to identify heterosexuality as the root of all women's oppression and to argue that separating from heteropatriarchal society was the only way of dealing with it. In order to avoid maintaining or perpetuating patriarchy in any way and to enable women to construct a new society beyond the influence of men, some lesbian feminists adopted the spatial strategy of distancing themselves from mainstream society by establishing separatist communities that excluded all men. Although some of these were established in urban areas, for example in Toronto, the aim of separatism was seen as best fulfilled in rural areas – because spatial isolation meant

Box 3.6 KATE

KATE IS A 37-YEAR-OLD BANK MANAGER. EACH LESBIAN RELATIONSHIP SHE
HAS BEEN IN HAS CHANGED THE WAY SHE EATS.

'It's odd 'cos I can remember in terms of a time when Helen [an ex-partner] was wanting
to lose weight and we kind of said that if that was going to happen, what we didn't want
to happen, 'cos I was living with her at the time, was we were cooking two separate meals,
'cos that's ludicrous, you can't get into that. So I kind of said, it's like I wasn't bothered,
you know, so there were periods when Helen was doing that that my diet changed in a way.
But it was really odd 'cos Helen was very successful in terms of when she wanted to lose
weight she did, and I was eating the same but I didn't lose weight. But I don't know why.

'I was with, she's one of my best mates now really, a woman called Catherine [an
ex-partner] who's a coeliac. So that was interesting, I was interested in it in a way, in what
she couldn't eat, and she has to look at everything, I mean she's got off pat now in terms
of what she can and can't eat. But it was interesting, I mean in terms of that period because
– what happened at first was I didn't dare cook for her [laughs] basically! . . . So I was like
tootling off to Chesterfield [where Catherine lived] all the time. But then it's obvious things,
like, in terms of things she can eat, then you soon discover and learn and adapt what you
cook when she came for a meal and stuff. So I'm aware of the problems, like we did the
coast to coast walk, four of us, and Catherine was one of them. And really stupid things
like she couldn't have toast in the Bed and Breakfast, and so she, like, carried these little
loaves of this bread which was just a bag of crumbs by the end! And, like, I went through
a period of, like, really adapting what I ate to her when she came round.

'For about seven years I didn't eat chicken, or any meat, just fish and then I got involved
with somebody, and I've always liked chicken! And it was so easy going back to it. Yeah,
I got involved with someone who is a meat eater, and so I started eating chicken again.

'The other negative thing to do with food is something, I've become really conscious of it
just recently, is about somebody [her new partner] trying to mother you with food, in that
their idea of taking care of you is feeding you. And it's driving me up the wall, I have to
say! . . . It's partly to do with my routine and time and everything like that, but it's also
partly to do with, I just can't face food in the mornings. Not, like, at seven o'clock, I can't
hack it. And, erm, what I've become aware of is the last long-term kind of relationship I
had, we had similar patterns in that sense, in that Helen didn't have breakfast really, you
know, coffee or whatever, and I'd have juice, but we didn't eat in the morning. And what
has happened now [with her new partner] is that I'm, like, force-fed in the morning, which
is a dreadful way of viewing it, but this is someone who has, who does enjoy food on a
whole different level, on a whole different plane, and breakfast is really, really important,
so she will always have breakfast . . . but like at seven o'clock in the morning I just can't
face it! I mean, it's really difficult in that situation, I mean, you know what it's like when
you're at the beginning and you kind of want to please [laughs], and I'm kind of trying to
eat this, and I've ended up saying now, "I really can't eat this", you know? . . . And that
is something – I've adapted to other people in other ways, like eating chicken, but being
able to eat in the morning is something I just cannot adapt to. And I never have adapted,
I've been in relationships with other people who eat breakfast regularly but it's not been
an issue, it's not something they've kind of forced on to me.'

that it was easier for women to be purer in their practices in the country than in the city, and because essentialist notions about women's closeness to nature meant that the countryside was identified as a female space. Self-sufficiency, particularly in relation to food, was important in these attempts to establish 'lesbian lands' (Faderman 1991; Bell and Valentine 1995b). Land trusts were set up to make land available to women for ever, and the separatists sought to create new ways of living and to work out new ways of relating to the environment by growing their own food in harmony with nature. Much energy was put into learning about herbs and 'natural' foods and developing outdoor skills.

Similar visions of ways of living, in which food preparation and consumption take place collectively rather than in 'family' units, characterise many feminist re-imaginings of non-sexist housing and home life. Hayden (1980), in speculating on what an alternative form of housing might be like, makes an argument for a new form of living in which private homes are designed and built around collective spaces and activity centres (including kitchens and allotments). Food co-operatives would grow food for the community's needs, and meals would be prepared and served not in the private homes but in these collective spaces.

Allotments are an economic necessity for some households, but for others they also represent a new lifestyle choice. Certainly in the 1990s there appears to be a greater awareness of green issues and of the possibilities for households to establish self-sufficiency in food to various degrees. For some, these domestic consumption practices represent an outright rejection of a capitalist lifestyle; for others, their lifestyle is motivated by concerns about the way in which commercial foods are processed, and represents an attempt to eat 'natural' products free from artificial, unsafe and unhealthy production methods (Haastrup 1992). Self-sufficiency also offers the added attractions of control over what you eat, the chance to see and consume the products of your own hard work, as well as the more abstract spiritual pleasures of working outdoors, as this person describes: 'Self-sufficient production cannot make enough money to cover the cost of the place, but it doesn't have to. We are not dependent on it. I do it because it makes me feel an inner beauty' (interviewee quoted in Haastrup 1992: 60).

Some households are only self-sufficient in certain foods. Haastrup cites the example of a household self-sufficient in apples. The problem with this growing strategy is that they cannot always control the quality or quantity of their produce and so apples appear in virtually every menu during the season, and are processed into different forms to be consumed at other times of the year as well as being swapped for other foodstuffs with friends, relatives and neighbours.

It is apparent from all the material discussed in this chapter that individual and household identities are produced, articulated and contested through food consumption and the spatial dynamics of cooking and eating at home. But some of the examples used in the chapter also highlight the interrelationships between sites of identity formation such as the body, work and school in these processes of negotiating identities at home. In the next chapter on community we pursue further some of the themes about collective eating raised in the last section of this chapter.

4

Community

ROBERT W. KATES (25)
GFW Xocalatl Truffles

●

Put 8 oz of very good bittersweet chocolate in the top of a double boiler over boiling water until melted.
Stir into the chocolate:
- 4 oz unsalted butter (melted over a low heat with foaming skimmed off);
- ¼ cup heavy cream (heated to a boil with three strips of orange peel, cooled, and orange strips removed);
- 3 tablespoons Grand Marnier or non-alcoholic liqueur.

Place in refrigerator to cool until thickened (15 or more minutes).

When thickened, form into ¾-inch pieces with two teaspoons and drop into baking dish lined with a very good cocoa.

Return to the refrigerator for 30–40 minutes until thoroughly thickened, then shape into a miniature globe with your fingers, roll in cocoa, and then put the resulting 30–50 truffles into appropriate wrappers.

Temper your delight in the truffles with your knowledge of the Colombian exchange, Aztec origins, bio-engineered BSH-free butter and cream, monastic agrisols, and geographer-chocoholics.

COMMUNITY

Community means different things to different people, and so it gets articulated through food in many different ways. This chapter seeks to explore these, always keeping sight of the fundamental theoretical questions which circulate around the notion of community – as well as those many commonsense ways in which we all think about it. We can instantly think of places where sharing food and drink helps bond us into a community – the local pub, or a street party, for example. But we must always be mindful of the fact that communities are about exclusion as well as inclusion; and food is one way in which boundaries get drawn, and insiders and outsiders distinguished.

There are also forms of community which coalesce in particular site-specific contexts. Institutional settings – schools, workplaces, prisons, hospitals – offer us further insights into this food–community equation. In such settings food can be both a form of control and a source of resistance, and 'communities' can be forged out of both situations.

Communities which transcend site specificity must also be considered, since routine practices and habits (including cooking and eating) are a common way to shore up community identity when geographical proximity recedes. Diasporic communities, for example, often maintain a sense of identity and history through food consumption. More broadly, in a world where consumption is so central to identity formation, the 'community of consumers' might supersede other aspects of identification, especially over particular campaigns and issues.

4

COMMUNITY

•

Community'. It's a word we all use, in many different ways, to talk about … what? About belonging and exclusion, about 'us' and 'them'. It's a commonsense thing, used in daily discussions, in countless associations, from 'care in the community' to the Community Hall; from 'community spirit' to the 'business community'. Soap operas and related TV genres are built upon structures and stories of community, from Albert Square to Brookside Close (sites of two UK soap operas), and from the bar in the US sitcom *Cheers* to Kevin Arnold's suburban neighbourhood in *The Wonder Years* (a nostalgic American serial about childhood in the 1960s). Many of us would also lay claim to belonging to at least one community, whether it is the 'lesbian and gay community' or just the 'local community' where we live. But, as was noted in the introduction, community is also one of those terms that is almost impossible to define, as it means so many different things, and this has vexed geographers and other social scientists who have sought to explore and explain how communities function. And, of course, geographers being geographers, numerous attempts have been made to create maps of communities, or at least to tie 'community' to location. Part of the use and understanding of community is spatial – the term can describe an area bounded in the minds of its occupants, and set apart from surrounding areas; thus any city might be fractured into a series of communities. But there are many more levels and scales to it – and a crucial point is that the term community is not only descriptive, but also normative and ideological: it carries a lot of baggage with it. Sociologists have tried to capture this many-sidedness, coining and re-evaluating terms like *Gemeinschaft*, or the *Bund*, to further specify concepts (Jary and Jary 1991). In many pioneering studies, urbanisation was seen to lead to a 'loss' of community; communities were idealised in rural areas, where small numbers of people interacted face to face (see, for example, Wirth 1938). Subsequent work, however, has emphasised the constrictions of such tight-knit communities, especially for those who do not 'fit' community norms (e.g. Kramer 1995; Philo 1992).

COMMUNITY CONUNDRUMS

Comparatively recent re-evaluations of what Jary and Jary (1991: 98) describe as 'one of the most difficult and controversial [concepts] in modern society' have sought to unknot some of the previous conundrums of community, and provide something closer to the commonsense understandings we all use. Worsley (1987), for example, proposes three broad meanings which are certainly usable here: community as *locality* – similar to neighbourhood, with common delineation of boundaries and a localised identity; community as a *network of interrelationships* – often losing spatiality, here we have what geographers sometimes call 'community without propinquity' (i.e. without physical proximity); identities are shared, but traverse space, as in the case of ethnic communities. Worsley's third meaning is an extension of the second, and specifies the *kinds of (or the qualities of) interrelationships* between members; this is embodied in notions like 'community spirit' or the 'sense of community' in a purely affirming sense. Of course, all these can and do overlap.

With an explicit focus on place and consumption, Robert Sack's (1992) discussion of community is particularly resonant for our work here, although his sentiments may not always echo our own. He argues very much in favour of local community, with its 'particular system of production, consumption and other social relationships [which] overlap and are enclosed within a single place. For the people living there, the place becomes their world' (Sack 1992: 188). As he says, such a 'territorial definition of social relations' is implicit in the structures of local politics, where an elected individual stands on behalf of her or his community (in the UK, local councillors embody this). However, given the profound changes we are all witnessing in the world around us, such localism may be failing adequately to represent our dominant allegiances and obligations; while Sack says that personal contacts outweigh anything the electronic media have to offer as the basis for creating community, affirming the centrality of place, he also presents arguments against local(ity)-based forms of community:

> It could be said that local communities smother their members in traditional obligations; that they stifle change; and though they foster responsibility and purpose, they make these moral precepts so narrow and provincial that everyone who is not a member is a stranger to whom the community has few obligations. Rootedness and locality become more virtuous than mobility and cosmopolitanism. In short, local community interferes with the dynamism of modern life and with the expansion of moral responsibility to include all human beings and even all of nature.
>
> Sack 1992: 190

These arguments, he goes on, are 'more dialectical than antithetical' (190); indeed, all kinds of rhetorics of community, often contradictory, are in circulation. Politicians currently obsessing over 'communitarianism', journalists hyping the 'global community' of cyberspace, queer activists attacking the exclusiveness of the 'gay community' – many commentators are struggling to make sense of what community means. Of course, it means all these things, and a lot of others. The debates rattle on, and it is perhaps time to disengage with them, and turn on the oven. Or, rather, to break out the bunting and start buttering bread for sandwiches, because nothing brings a community together like a street party! (or so they say) . . .

On 8 May 1995, Britain was overtaken by VE Day celebrations; aside from the huge jamboree in Hyde Park, the most prominent features of the time were street parties. Following a tradition reputedly begun in Britain with the signing of the Treaty of Versailles in 1919, but now more commonly associated with royal occasions (the most memorable recent ones being for the Queen's Silver Jubilee in 1977, and for Charles and Di's wedding in 1981), the VE Day parties typically used a mix of 1940s and 1990s iconography: wartime singing legend Vera Lynn and 1990s kids' TV character Mr Blobby, karaoke and tea dances; a party at Stedham in West Sussex blended wartime rations and cookery with Thai food donated by a local restaurant (Lacey 1995).

But are street parties cause for everyone to celebrate, and for local communities truly to come together in unity? A case like VE Day is, of course, immediately contestable, since it marks a nationalistic moment, as the ongoing arguments about the inclusion of German ex-servicemen in the Hyde Park jamboree showed. Closer to home, not everyone wants to celebrate the British war machine, or the Second World War. Just as the Silver Jubilee and royal wedding were criticised by anti-royalists, so VE Day need not unite everyone (despite the most romantic portrayal of community togetherness in the best-attended street party of May 1995 – that thrown by the residents of Albert Square in the BBC's soap opera *East-Enders* and shared with millions of viewers). But apart from these ideological challenges to the homogenising thrust of street partying, the everyday pros and cons of 'community' articulated in butties and bunting equally attract some and repel other people:

> The idea of an alfresco frolic with the neighbours evokes both cheers and cringes. 'It's wonderful, really brings the community together,' says Mrs Wendy Keeble of Hornsea Road, Portsmouth, who has galvanised her locale into action.
>
> Young householders, however, tend to be rather sniffy about the whole idea. 'A street party is such a tacky notion,' says one from South London. 'I remember the Jubilee party when I was still living with my parents. It was embarrassing – all these crusty ratbags whose normal communications are complaints suddenly flinging decorum to the winds and bounding about getting pissed in public. I don't even know my neighbours.'

So many events are being planned that party-haters are having to go to considerable lengths to escape. 'We can't think of an excuse to get out of our neighbour's VE Day barbecue, so we're leaving the country for a long weekend and going to Ireland,' says one couple. 'We don't like them, and we don't want to eat their sausages.'

Lacey 1995: 22

A second recent example serves as an interesting contrast here. A month after VE Day, British church leaders, always prominent users of community rhetoric, welcomed 300 guests to a grand banquet in Whitehall, London:

The Leader of the Opposition took supper with a drug addict last night. Lara Newton learnt about political abuse, Tony Blair about substance abuse. A firefighter and a homeless person seated at the same table listened in. Nearby, John Birt, Director-General of the BBC, was chewing the cud with a train driver, another drug addict and a Roman Catholic Cardinal.

Wroe 1995: 2

In effect only as forced as a VE Day party, the church's community-making great banquet attempts to solve today's 'social problems' through commensality and table-talk. In the 1960s, the Soviet intelligentsia evolved an informal, salon-like 'kitchen culture' for discussion, flirtation and the cementing of friendships over a meal; this unofficial, unpredictable social space contrasted with the bureaucratised lives of many Soviet citizens at the time – including those not fortunate enough to live in apartments with their own kitchens, who had to endure the rigidly formalised sharing of communal kitchen space (Boym 1994).

Communal eating has not always been thought of as good social work, however. The increasing domestication and privatisation of eating has been seen by historians as a way of securing against rebellion forged over a shared meal (Driver 1983; Lang 1986/7). In the same way, working men's clubs were started as a better alternative for the rational recreation of workers than uncontrolled alehouses, although they too eventually let alcohol in, as Williamson (1982) explains in his study of a north-east England mining community. The importance of drinking – and more especially of drinking-places – for community solidarity is discussed in some detail below, as are the possibilities for both discipline and resistance in other communal eating environments, such as works canteens and refectories.

But first, let us return to Bill Williamson's (1982) *Class, Culture and Community*, which uses techniques of biographical sociology to build an account of social change in a mining community through a detailed study of one man's life, that man being the author's grandfather, James Brown (1872–1965), Northumberland pitman. Here, in the early part of this century, in an old mono-industrial northern English heartland, we find an archetypal form of community: village centred, class bounded, work related, close and closed. In a chapter on 'time off', Williamson outlines the pit families' leisure activities (or, to be more accurate, the *pitmen's* activities, since many of these were gender and age segregated). Aside from hearth and home (where he sat and smoked his pipe), Brown enjoyed drinking in the working men's club and local pubs (but *never* to excess), gardening (both at home and on his three colliery-provided allotments) and, especially, growing and showing prize leeks and keeping pigs. A mixture of self-sufficiency, 'rational recreation', tradition and a way of securing status position within the community (of which Brown was apparently ever-conscious) motivated him in the hard work of leisure. These kinds of activities wove a web of social contacts, whether the mild rivalry of leek growing, or the neighbours saving scraps to feed the pigs and assisting in their butchering (for which they were rewarded with sausages, black pudding and broth), or the singsongs at the pub. Everything, and everyone, was interlinked.

Williamson's description of drinking in the social lives of miners is particularly interesting. He says of his grandfather:

> [D]rinking for him was not a thoughtless indulgence, expensive or even ruinous to his family. . . . Quite the opposite: it was a strictly controlled activity, almost solemn, and he took pride in being able to 'take a drink' sensibly, distinguishing himself and his friends from the less respectable boozers in the district. For him, being able to take a drink properly was a small but significant part of his sense of his social status.
>
> Williamson 1982: 106

Brown disapproved of women in pubs, and so was keen to transfer his allegiance to the male environment of the working men's club upon its opening in 1908 (indeed, he was a founder member of the club). Open all day until 1914, the club was a centre of (male) community life. While this is celebrated in Williamson's account, other commentators have suggested that the social cohesion offered by pub life is exclusive and can work to maintain social barriers. Hunt and Satterlee (1986), for example, looked at a village undergoing the transformations of rural society at the end of the twentieth century. Affected by counterurbanisation, the village in their study, Melton in Cambridgeshire, has seen its population rise from 1,600 in 1951 to 3,892 in 1981. A perceived 'loss of community' has resulted, with conflicts between 'established locals' and 'newcomers'. Nowhere is this felt more strongly than in the village pubs, where in-groups of 'regulars' are tightly bounded, enhancing the sense of identity for group members, but also emphasising the exclusion of others, often referred to as 'outsiders'. Drinking protocols include many techniques to keep outsiders out, called 'freezing' by Thomas (1978). As Hunt and Satterlee (1986: 525) note, in Melton 'pubs operate as important cohesive institutions', but this cohesion is 'counteracted by the splintering of village life into discrete social groupings, which use particular pubs as their meeting places'. The unfolding drinking dramas of the two Melton pubs reveal how social structures – most notably class – get articulated through socialising in particular ways in particular pubs, and the importance of key drinking rituals, such as the buying of rounds, in maintaining a sense of 'community' among drinkers (while also marking the buyers' standing). The social worlds of the two pubs, moreover, rarely interact, despite the fact that both are involved in essentially the same activity – drinking – in the same place. In Box 4.1, John talks about the different pubs, each with a different sense of community, which he visits on different nights of the week, describing the social awkwardness he sometimes feels when he is on the boundary between being an insider and being an outsider.

A study by Gerald Mars (1987) of the drinking cultures of longshoremen in Newfoundland provides a further example of how cohesion and division, insiderness and outsiderness, become articulated through drinking practices. In the same way as Holliday's (1995) study of 'core' and 'periphery' workers in small firms revealed differences (in treatment, perception, behaviour, and so on) between people doing the same or similar jobs in the same workplace, Mars' study shows that 'core' workers (or 'regular men') maintained a close-knit circle which expressed itself in out-of-work social situations:

> There are marked differences in the way these two classes of longshoremen drink. These can be seen in the rate they drink, the locale they occupy and even the beverages they choose. Regular men drink in taverns close to their wharf; they sit with their regular workmates and drink beer: outside men, even though working on the same wharves, usually drink in the open air or sit in parked cars and they drink rum or wine in smaller groups that vary in membership.
>
> Mars 1987: 91

The drinking rituals reflect the patterns of work, especially the system of hiring of work gangs and the loyalties which develop between regular men in gangs; in fact, more than reflecting these work patterns, drinking draws an explicit link with the rhythms of working life on the waterfront. Those who work together drink together – as long as they're regular men. The men don't buy rounds, but do buy drinks for a couple of their companions, with core members buying fewer drinks than those trying to establish themselves within a group. For 'outside men', tavern drinking does not fulfil the service of repaying obligations for those whose work patterns are more erratic (outside men often fill in for regulars who are off work; this temping means they never become

Box 4.1 JOHN

JOHN IS A RETIRED UNIVERSITY LECTURER. HE IS MARRIED AND HAS TWO SONS. HE HAS A REGULAR ROUTINE OF GOING FOR AN EARLY-EVENING DRINK, VISITING DIFFERENT PUBS – EACH WITH A DIFFERENT SENSE OF COMMUNITY – ON DIFFERENT NIGHTS OF THE WEEK.

'Saturday and Sunday are university people that I've known for many years – different people, I've known them for many years. Wednesday, when I go to the Black Horse, that tends to be university people. Friday, it tends to be university people when we've been to the cricket. But Monday, Tuesday and Thursday, it's the, it's a business crowd really and it's quite a different perspective on life and they're in the real world sort of thing, spending £44,000 on cars and this sort of thing.

'I'm an outsider with this business group at the White Lion, I don't think I'll ever be a full member of that group, for one very good reason is that one of the things that I've been involved in over the years is buying rounds of drinks and this is a prescription for drinking too much, there's no doubt about that, in my view. Anyway, with this group I didn't want to get involved in buying rounds, that was the problem, and I think that if you are a part of a group, it's always very awkward when I take off and go and get myself a drink and then I come back and join them again. It's awkward for them, because they will count round, they'll go round the group, you see, and they miss me out, you see, and it's awkward – you can see some hesitating, someone will ask, and so it does make life awkward. But I didn't want to get involved in buying rounds because some of their rounds could be, well, easily £10, which is far too much for me to spend in one night. So I shall never become an insider from that point of view, but I suppose I'm not an insider because I'm the only academic there. In the university groups that I mix with now, apart from the small group that meets on Sunday night, we tend to, we tend to share rounds. If about eight people get together then we'll – and we've eight drinks, let's say, well two of us will buy the drinks.

'And it is, kind of, competition when you're buying rounds, you know, you don't want to miss out, but I mean the rounds now are so expensive – their rounds are – and they're quite likely to go on to what they call a top shelf, which is whisky, and so I find it too expensive.'

members of any particular work gang). Instead, cheap wine or rum is consumed somewhat furtively in the open (outdoors drinking is illegal in Newfoundland, so the drinkers are ever watchful for police, who know the longshoremen's reputation for open-air boozing) in a flux of loose social groups controlled by the 'hosts' – the men with the bottles. In this way, quite distinct drinking rituals maintain the two groups, and maintain their separation.

Other work-based food rituals similarly reproduce the social order of 'work communities'. Pitman James Brown was a regular attender of the colliery picnic, which was a key event in the work's social calendar (Williamson 1982). Outings, parties, festive dinners and picnics are thrown by many organisations formally (added to this list are all the forms of informal socialising between co-workers), and are often seen as important for forging a sense of community among colleagues, and between bosses and workers. An account of picnic day on the campus of the University of California, Davis – in part a hangover from its origins as an agricultural college – describes the parade of floats and bands as 'a fine folk instance of a temporal, linear syndetic work – an anciently legitimate kind of open-ended community work' (Mechling and Wilson 1988: 306). Such events, they suggest, 'become occasions for shared, public discourse (both verbal and nonverbal) "about" troubling ambiguities and contradictions in the organization's symbolic categories, norms and values' (316); by allowing some (constrained) transgression, employees can 'let off steam' in a ritualised, sanctioned, 'safe' way, thus dissipating tensions. Maintaining control, however, can be difficult, and transgression and tensions may escalate; outbreaks of fighting at works parties are not that uncommon, and the careful management and surveillance of transgression marks such events as symptomatic of 'controlled de-control' (Rosen 1988).

Communities formed at work, or in other institutional or organisational settings – schools, hospitals, prisons – use food and eating in numerous ways. Food can be a form of resistance, a form of discipline, of reward, a way of creating 'community' or a way of refusing or denying it (Holliday 1996). As Sallie Westwood's respondents show in *All Day Every Day* (1984), for women working in factories, opportunities for resistance can take unlikely food-related forms: a birthday means cake, which means time off the machines to eat it which the bosses won't interfere with (no one wants to be a party-pooper – although when one woman overstepped the mark, getting drunk at lunchtime on her birthday, she was subsequently fired). Concerns over the health of employees, schoolchildren or inmates promote regimes of 'healthy eating', free cholesterol checks and so on. At work, taking care of employees has become a mainstay of the new human resources management, echoing earlier paternalistic concerns enacted though shows of benevolence such as the works outing (e.g. Delgado 1977; Townley 1994). But these 'benevolent' caring strategies can create communities of resistance responding to 'health fascism' and the erasure of fundamental human rights – such as the right to smoke. No-smoking policies in the workplace make outcasts of smokers, who have to huddle outside (businesses are even beginning to worry about the image they lend to the organisation, making them smoke in invisible, backstage hideaways; see Anthony and Miller 1995; Klein 1993). And, of course, scandals about companies pressuring overweight staff to diet – or just sacking them – show how health promotion and corporate image are intertwined to the point of confusion (McDowell 1995).

In certain work cultures, the provision of food for employees takes on a symbolic role. For some large corporations, for instance, the 'company breakfast' has become an important work/social occasion, a place to celebrate and reward commitment and excellence through the commensality of the (updated) feast (Rosen 1985). The symbolism of food at work may be much more marked than this, however. In the logging companies of the USA in the early part of the twentieth century, standards of food provision became central to both company image and employee loyalty (Conlin 1979). Working and living in incredibly primitive conditions, for long hours, with low wages, lumberjacks developed a working culture which commanded high standards of culinary care. The hard physical work, it was estimated, left loggers needing to consume up to 9,000 calories a day, and their food intake was of gargantuan proportions:

> In West Virginia in 1907, forty-five men stowed away in one week: a tub of lard, a sack of turnips, a sack of onions, a box of yeast, a case of cream, a barrel of sweet potatoes, 7 sacks of Irish potatoes, a case each of pears and peaches, 2 cases of eggs, a case of tomatoes, a barrel of apples, 112 pounds of cabbage, a case of corn, 22 pounds of cakes, 10 pounds of tea, 12 cases of strawberries, 2 barrels of flour, 15 cans of baking powder, and 300 pounds of beef.
>
> Conlin 1979: 168

Such a privileged diet became seen as the norm for loggers, and every innovation in food technology (preservation, new foods and so on) was demanded as soon as it became available. Gourmet menus were eaten by men who put up with lice, lack of sanitation, cruel weather, low pay, and who showed little interest in unionisation. And when the labour market began to tighten, it was on food provision rather than wages or other working conditions that logging companies competed for men; companies put up menus to tempt workers from rival camps, and professional cooks replaced loggers' cooking by turns. Camp cooks were among the most respected employees, commanding wages similar to those of foremen, and being allowed considerable eccentricities. They also commanded silence at meal times, and loggers commonly sat and ate their way through a 3,000-calorie meal in less than twenty minutes. Comparable situations are described by Orbach (1977) among tuna fishermen around San Diego – skippers would brag over the radio to each other about their ship's cook and the meals they provided – and in Johnson's (1979) study of a fishing community in the Algarve, where communal meals at the end of a trip were a way of resolving tensions among crew members.

EATING IN INSTITUTIONAL COMMUNITIES: CONTROL AND RESISTANCE

Not everyone in an organisational or institutional community, of course, can command such power over food; in many cases, in schools (Plate 4.1), prisons and hospitals, for

Plate 4.1 'To serve them all our days'
Photograph: Gill Valentine

example, food is provided with minimal (or no) consent and its consumption is heavily policed. Indeed, the routines and tastes established in these institutional communities may well continue to influence some residents' eating habits long after they have moved on to other places; while for others the memories that the smell or taste of institutional food brings back are enough to put them off certain meals for life.

School dinners were initially provided as a way of countering the poverty of many children's diets at home, by ensuring a universal meal for every schoolchild (Rose and Falconer 1992; Berger 1990). But in contemporary British schools, compulsory competitive tendering and the local management of schools have transformed traditional school catering. Now discourses of consumer choice and competition dominate. Morrison (1995) describes a co-educational secondary school for 11- to 19-year-olds. Here, the pupils have a choice from a range of 'healthy' and 'junk', and hot and cold foods in the school canteen, as well as vending machines on the school site. The food practices in this institutional community, however, owe as much to the school's need to spatially control and organise its pupils as they do to concerns about children's nutritional welfare. The proximity of the school to the town means that at one stage children were pretending to go home for lunch but were actually 'skiving off' into town to eat 'junk' food. Hanging around the streets and occasional incidents of shoplifting began to erode the image of the school and so it attempted to reproduce the attractions of the town on-site. As Morrison (1995: 243) explains, 'Institutionalised snacking is, in major part, an organisational strategy which solves the daily management problems of controlling pupils.' Eating in this institutional community begins as early as 7.30 am when pupils begin to form queues round the vending machines. The spatial limitations of the dining-room also mean that the eating culture of this institutional community is characterised by cafeteria-style queuing and rapid eating in place of the traditional 'family service' system which characterised school meals until quite recently. Many pupils adapt to this by abandoning any pretence of table manners, using their fingers to eat their food.

Within the broader 'community' of the school, food also plays a part in constructing 'communities' of year groups and 'communities' of friends. At lunchtime and break times the pupils divide into social groups to eat lunch and share snacks and drinks. In this way the choice and the sharing of food can define 'insiders' and 'outsiders' along the lines of gender, ethnicity, income, age and body shape. While drinking alcohol is important for boys to be socially included, girls worry about being stigmatised for being overweight (Morrison 1995).

Like eating at school, collective eating in the institutional community of the prison is very much constructed around issues of control. Mennell, Murcott and van Otterloo (1992) describe this sort of eating community where 'members' have limited autonomy (hospitals are another) as a 'total institution'. The prison officer, in Box 4.2, describes how food is central to the management of his prison community. In this particular institution the inmates are not allowed any choice at meal times because of a fear that it may lead to bullying and violence. The timing of the meals is dictated by the work routines and rotas of the staff. Much to the irritation of the prisoners, this means that their food is served much earlier than they would normally eat it at home. In this way the institutional routines of this community dictate the body clock of the inmates.

All food is served on the wings from trolleys. Having collected a meal, inmates are locked in their cells, and for some this means being forced to eat alongside their toilet or slop bucket. Some choice is available in the prison from the tuck shop, which serves a range of confectionery, snack foods, fruit, as well as cigarettes and phone cards. The shopping habits, like the meal habits, of the inmates are another way in which the staff keep an eye on what is going on in this institutional community. Rapid selling of certain items is usually an indicator of some form of trading related to the gang or drug 'communities' on particular wings.

But institutional eating also offers opportunities for subversion and rebellion for people who have seemingly little control over their lives. For example, prisoners

Box 4.2 MR HIGGINS

MR HIGGINS IS AN OFFICER IN A MEN'S PRISON. HE BELIEVES FOOD PLAYS AN IMPORTANT ROLE IN THIS INSTITUTIONAL COMMUNITY.

'Yeah, yeah, the biggest flashpoint in any prison is down to the food. If the food is total crap you have got trouble. You can usually tell, most prisons that go, it's down to the food. If you can serve decent quality with a substantial portion, the lid'll stay on. If you start giving, oh well, you know ... we'll give a few canapés of caviar, the roof'll come off, 'cos it's not just quality they want, they want quantity. So you've gotta try and blend it all in to get a happy medium, because no matter what you do, you won't please them all. But if you can keep the majority happy, it keeps the roof on your prison, 'cos it is a flashpoint.

'The reason there's no choice.... Until you can organise it so everybody got a fair choice, there'd be trouble, because they're inmates. And Joe Bloggs coming down halfway through when everybody's eaten the gammon, he'll be saying, "I want a bloody gammon steak – he's had a bloody gammon, why can't I have one?" And the next thing, you've got fighting. I've seen them fighting on a hot-plate over a size of a sausage – one guy's sausage had been longer than the other person's sausage and: "His sausage is bigger than my sausage", and they'll fight over petty things like that, there's nobody to back us up – they're so petty – the blob of cream on top of the trifle can cause the trouble, 'cos he's got more cream and jam on the top of his trifle. So if you give everybody the same choice, all get the same, plus it stops the bullying.

'If somebody's off their food – if a person refuses food, there's an automatic – straight away – it's these lads do it – he hasn't had his food, or a prisoner says: "I'm not eating" – what is it ... 10.15 ... a refusal of food form, fill it in and you monitor it. Then you've got to offer them it ... If he doesn't – he's got, if he wants to say, "I'm on a food protest", he has gotta go through it properly, he's gotta go down to the hot-plate, collect a meal, take it in his cell, keep it in his cell and then an hour later you've to go and get it out – that is how he does it, otherwise it's not gone, it's not a valid food strike, if you like.... And then if he carries on he's then taken down the hospital where he's seriously monitored there. It's down to jailcraft again – I remember two years ago [a governor] ran one down to the hospital, he hadn't eaten for weeks, and this governor came down to me and said: "Have you got some oranges?" So I said, "Aye", I said, "What do you want oranges for?" "It's an old trick", he says, "I learnt." He got a half a dozen oranges and he just chopped them up and he says: "Give us a tray." And when you give them a meal you put the oranges on, and in twenty-four hours the guy was eating. He says, "It's an old trick we learnt – if they don't eat, chop some oranges up, put them in the cell and you can guarantee nearly 90 per cent will start eating again", and it works – that's jailcraft – an old thing he learnt thirty or fifty years ago as a young hospital officer. So – and I didn't have a clue, and he had some oranges and chops them up ... and nobody worries if prisoners won't eat.'

sometimes choose to protest their innocence while in prison by going on hunger strike, even to the point of death; and religious cults are renowned for their use of diet control as part of members' 'programming' (of course, food has immense symbolic importance to many religions; on its importance for enculturation and religious conversion in US Krishna communities, see Singer 1984).

Steve and Mick, two men shortly due for release from prison, explain in Box 4.3 some of the ingenious strategies which they or other inmates have employed to resist the institutional diet of the prison. In the first quotation Steve describes how some of the Muslim prisoners, unhappy with the British and Americanised foods on offer, will hide some of the meat which they are served and then later try to heat this up in their cells, adding herbs or spices smuggled into the prison by their friends and relatives. For many inmates food can be particularly evocative of 'home'. But while those in other total-institutional communities, such as boarding schools and hospitals, are often brought home-made foods by visiting friends and relatives, which give them some sense of

Box 4.3 STEVE AND MICK

STEVE AND MICK ARE BOTH NEARING THE END OF THEIR TERMS IN A MEN'S PRISON. THEY HAVE BOTH DEVELOPED STRATEGIES FOR RESISTING THE INSTITUTIONAL DIET OF THE PRISON.

'. . . a lot of them do – even save burgers, anything, you know, like someone'll save a cold burger, you know, with bread – I know it doesn't sound nice, but just so they've got sommat to eat, put sauce on it, whatever . . . and what it was, kids was pinching sheets and putting oil on it or whatever, and they burn it in the cell and they put food on the tray with a bit of butter and try and cook it up, don't they, in their cell, and that's what they're doing.'

'Oh yeah, Marvel tins and that – you know what I mean – you put string through it and hang it on the back of your chair and boil it with a piece of sheet for your cup of tea and that like. 'Cos, like, there's only [one of the prison wings] that's got kettles in cells – do you know what I mean? – so the rest don't have them.'

'Well [two of the prison wings], they face each other – right – and my mate were on [one of those wings], and I could throw a line – don't forget we're talking like twenty or thirty foot – I can throw a line from my window – it landed on some cage, and he threw one and it tied with mine and he pulled mine in, and I tied him a bag of food down we made like a pulley, and people could wonder, used to wonder how I got, how it'd get, you know, like across from one side to another. I mean, it's easy to open a window, just put your arm and swing it across – you do that now – pass sandwiches to each other – tie it with a bit of string and he puts a stick out and catches them. You'd be surprised what they come up with, innit? Just a little thing in here, though, can make your day, can't it?'

individuality, autonomy and a taste of the outside (Crotty 1988), for security reasons prisoners are allowed not such luxuries. Food fantasies are therefore as common in prisons as sexual fantasies.

The subcultural community in the prison of those who regularly work-out in the gym reflects another way in which some inmates rebel against the institution they are in. Being fed three 'proper meals' a day, often more than most men would consume outside, means that those who take little exercise often put on a lot of weight while they are serving their sentences. For a small 'community' of prisoners, exercising in the gym at every opportunity and manipulating the prison food system by trading foods or bullying other inmates to obtain larger portions enables some prisoners to develop their physiques and thus resist the way the institution would otherwise reshape their bodies.

FOOD AND COMMENSALITY

In many of the more affirming portrayals of community, certain institutions are seen as cornerstones. From the perspective of food, we include here the pubs and clubs already mentioned, plus corner shops and cafés, as well as particular kinds of food festivals, such as street parties and works outings, to which we would add barbecues and other similar communal outdoor events ('pig roasts', summer fêtes and garden parties, for example), and countless other seasonal or religious events – Bonfire Night, Hallowe'en, Ramadan, Thanksgiving, harvest festivals, and so on. In all these settings, food retains its 'communicative' role, purportedly lost in the anomie of the hypermarket.

The centrality of these loci of food consumption for community-building and cohesion is echoed in their regular appearance in popular culture representations. In the Australian soap *Neighbours*, for example, the 'barbie' is almost always sizzling (it features, in fact, in the opening credits, as a device to show cast members together); the outrageous Hallowe'en lodge party is an annual mainstay of the US sitcom *Roseanne*, with community members trying to outgross one another with gory costumes and pranks; *Ellen*'s bookstore-cum-coffee shop Buy The Book is a nexus for the action in a further US comedy; and British soap *EastEnders* features or has featured a cafe, bistro, wine bar, fruit and veg stall, catering business and sandwich round, fish and chip shop, supermarket, allotments, the Queen Vic public house, as well as countless sporadic food events like street parties, wedding receptions and Christmas dinners – all of them occasions to forget differences and cook up some community spirit.

Food sharing, or commensality, as enacted in *EastEnders* at Christmas (when lonely old ladies are invited to join in extended family meals), is a great signifier of community, and anthropologists have emphasised its role in kinship and reciprocity ties in countless cultural settings. In Box 4.4, Walter, a widower who lives in a small neighbourhood of

Box 4.4 WALTER

WALTER IS 66 AND RECENTLY WIDOWED. HE IS RETIRED AND LIVES ALONE,
BUT HAS A STRONG SENSE OF NEIGHBOURHOOD COMMUNITY.

'I mean, the lady who lives next door to me, Enid – lovely lady, she's an old pastry cook,
and I mean she'll often bake something and she'll knock on the door: "Here, I've done you
a bit of this", you know, and it's all fresh-baked – parkin and things like that. Not a lot,
but while she's doing it she does me some. Stan in the corner says to Enid this morning while
I was talking to her, "Are you ready for going, love?" And they took her up to Kwik Save
to get her groceries in, you see. But equally so, she'll be going up to the bakery shop here
and she'll knock on door: "Do you want anything from bakehouse?" you see. And equally
so, I was going up to Asda last week and she wasn't doing anything and I says, "I'm going
up to Asda, do you want to come with me?" I thought, "Do you want to come on run and
just have a look round?" She says, "I've just got my groceries this morning." But I dare say
she would have, she would have gone with me. Now that is something here again which
is worth a lot, it is worth a lot – the old syndrome of the terraced housing had a lot to be
said for it, we pass one another's door, so we all speak: "Are you all right?"

'The thing was the jubilees and the last one I remember was the Silver Jubilee of the Queen.
They had the procession and we went to a street party that day, down in the little village,
it were, and then at night they'd got it all to have a celebration and a bonfire and they'd
got a big pig on a stick in the middle of the main street and you could all buy it. All the
houses took part, they all provided something. I mean, it was as if you was to have a party
with friends and you said, "bring a bottle."'

terraced housing in Sheffield nicknamed 'the village', describes how food plays an
important part in articulating the sense of community which exists between the residents.
As Mennell, Murcott and van Otterloo (1992) conclude, however, it is important to
remember the exclusions as much as the inclusions of commensality (soap opera
utopianism notwithstanding). Nevertheless, the romanticisation of commensality is a
powerful rhetoric. An advert run by the former new town of Milton Keynes appeals to
the local-community-forging capabilities of that most typical everyday form of commensality: local people drinking in the local pub. The advert's headline is just that word:
Locals (Plate 4.2). In an attempt to show olde-worlde unity in 1990s diversity, we see
photographs like that of Mac Makino, MD of New Wave Logistics (UK) Ltd, who
'enjoys a quiet drink with his wife, Nobuko at The Thatched Inn, Adstock'. Places like
The Thatched Inn are, the ad says, '[t]he focus of the community; a "drop in, see a friend"
sort of place'. 'No wonder', it adds later, 'the lucky locals who have chosen to relocate
their businesses to Milton Keynes seem pleased with themselves.' More famous for its

Plate 4.2 Olde-worlde unity in 1990s diversity
Source: *Good Housekeeping.* © CNT Marketing

concrete cows and (sub)urban anomie, Milton Keynes is here rebuilt as a community of *locals* (both the people and the pubs) through place promotion, cemented by pub commensality (a reference back to Hunt and Satterlee (1986) seems more than a little appropriate here, with a reminder about *exclusion* as the flipside of the inclusion advertised – we see MDs, CEOs and general managers sharing a pint of beer, but where are their employees? Or Milton Keynes' jobless?).

Food consumption has in itself the power to create 'communities' beyond the local, beyond the effects of commensality. Shared food habits also bind people together in what Brown and Mussell (1984: 11) label 'communities of affiliation'. Sometimes their combined consumer pressure is brought to bear on food producers and providers. Vegetarianism is a useful illustration, as anyone who has watched supermarkets responding to a rise in the numbers of customers eating a meat-free diet will know (as will anyone who has witnessed the meat industry try to tempt people back with a variety of advertising campaigns such as the recent 'food of love' TV commercials made by the British meat and livestock trade). In a countercultural milieu which could be seen as a form of community without propinquity, animal welfare, health, anticonsumerism, ecology, and concerns over world hunger come together to create a modern form of 'ethical eating' – veggie, as it is now colloquially known – whose diverse origins stretch back at least to ancient Greece (for an excellent review of the sociology of the 'vegetarian option', see Beardsworth and Keil 1992).

Vegetarianism (and all its cousins, from the anti-red meat 'almost vegetarian' to vegans and fruitarians) is only one form of ethical food consumerism, of course. The campaign against the food giant Nestlé for its cynical and manipulative marketing of infant milk formula in developing countries is a further notable example which has had widespread support (Chetley 1986), as is continued protest against McDonald's on health, humanitarian and environmental grounds which rallied an 'anti-birthday party' when 'Ronald McDonald' turned 40 (Carey 1995). This anti-McDonald's activism certainly flies in the face of the company's attempts to sell itself as community-maker through 'heart-sell' advertising evoking love, family and belonging, all under the glow of the golden arches (Helmer 1992).

Warren Belasco's (1993) *Appetite for Change* charts the making of a 'countercuisine' since the 1960s in the USA; he cites books like Frances Lappé's *Diet for a Small Planet* (1971), True Light Beaver's *Eat, Fast, Feast* (1972) or Ita Jones' *The Grubbag: an underground cookbook* (1971) as central to the emerging radical food consciousness, which hated fast food and celebrated 'slow food', hated 'white' (flour, sugar, bread) and loved 'brown', craved 'ethnic food' and loathed 'WASP food', and praised organic, small-scale, 'folk' food production against agribusiness. Rural communes were established to reclaim farming from transnational corporations and rediscover traditional production techniques – as well as giving a more united, affirming 'community' focus to the countercuisine: 'Being a "health food nut" in the family dining room or school cafeteria conferred identity, but it could be a lonely undertaking. In commune life one went beyond personal protest to build a cohesive model community' (Belasco 1993: 76).

As contemporary commentators attest, food was a more significant part of the communal project than the 'free love' and drug experimentation with which such projects are more commonly associated; the kitchen was more of a social centre than the bedroom, and the ideological connotations of every foodstuff were carefully considered, their implications in systems of oppression scrutinised from every angle. Food became the focus for intense soul-searching, and issues such as accepting government food stamps became the source of considerable anxiety. As one commune farmer, Elaine Sundancer, said, 'I want good vibes in my food, but I'm eating Vietnam for breakfast' (quoted in Belasco 1993: 80). Communities struggled with deep political differences, and sometimes fell apart as a consequence.

For those who preferred to stay on in the city and fight oppression from within, food co-ops sprang up as 'neighborhood outposts of the countercuisine' (Belasco 1993: 87), often getting supplies from farming communes. Originally called 'food conspiracies', they also carried information and provided a meeting point for urban radicals. Importantly, they arose out of a need to fight the forces of corporate consumerism in an era when activists began exposing the policies of supermarket chains, providing detailed product labelling and even counselling for shoppers on how to buy ethically. However, here too politics could cause divisions: if the co-ops wanted to attract a 'working-class' clientele (and in their revolutionary vision, some co-op workers saw this as essential), then they had to compromise on the price and range of produce they stocked, which did not please those whose agenda was wholly ecological. By the mid-1970s, most co-ops had given up on uniting the working classes, and settled for feeding the health conscious and eco-friendly. Somewhat inevitably, co-ops had to become more professional, more organised – ushering in an era of what Belasco calls 'hip business'. An important development which suited the move towards more comfortable forms of rebellion was the growth of 'hip restaurants', which

> attempted to institutionalize the communal [and co-op] culture. If food was a medium for 'communication,' much of the deviant dialogue was first carried out privately in communes and dorms, and then, as the first generation of converts aged and had more to spend, in public commercial spaces ... [their] loving attention, homey fare, and communal atmosphere helped diners battle everyday 'alienation.'
>
> Belasco 1993: 94

And so, via familiar forces of both social emulation and corporate poaching, through the 1970s health foods, vegetarianism and the whole countercuisine spread outwards but also became colonised by big business, while radicals became business people and aimed high; again, such culture clash left many people feeling exploited or disillusioned, since 'the oppositional grammar lost its context and power' (Belasco 1993: 108) as hip business became big business and health foods appeared on supermarket shelves. Nevertheless, the story of the American countercuisine is a particularly appropriate example of the mobilising rhetorics of 'food' and 'community', and how the complexities of both exist in constant tension rather than idyllic harmony.

The contemporary food consumer is faced with similar tensions: a recent hip new-journalism piece from US magazine *Interview* cites current trendier-than-trendy New York eateries catering a 1990s version of countercuisine, but has to begin by defending vegetarianism in the wake of something of a meat backlash (Rubenstein 1992). In Britain, the food producers have tried to convince the public to come back to meat after a series of scandals rocked their customer base. First 'mad cow disease' (BSE) ate into beef sales (already hit, along with other meat products, by exposés about steroids, hormones, factory farming, radioactive fallout and countless contaminations and adulterations), and now the activist focus is on the transport and export of live animals – especially veal calves. Here, once more, we find powerful community cohesion around food issues. In quiet seaside towns, local people have formed protest groups, chaining themselves to lorries and hurling abuse at livestock exporters. Throughout 1994 and 1995, pitched battles at ports and airstrips were matched by attacks on farms, the homes of veal farmers, and the government. Added to the scenes of action at places like Brightlingsea, across the country consumers rethought their shopping and eating habits, shunning certain products and producers (who have not been slow to react, offering food with ethical guarantees). If one can talk of 'communities of consumers', then one of the most active presently has set itself the task of scrutinising every link in the human food chain, boycotting those who cannot meet its standards and rewarding those who can with customer loyalty – and increased sales (Smith 1990). It is no longer enough for sellers to put glib slogans on labels like 'environmentally friendly' or 'cruelty free'; customers now want details.

At the time of writing, a news story broke over Freedom Food, which operates under the auspices of the UK's Royal Society for the Prevention of Cruelty to Animals (RSPCA) to improve conditions for animals farmed for human consumption, monitoring their care from birth to slaughter. Produce from animals kept in accordance with their guidelines is then sold with a badge of quality emblazoned on it. However, three specific practices permitted by Freedom Food – the trimming of chickens' beaks, the use of farrowing crates for pregnant sows, and tail-docking of pigs – have been highlighted as an RSPCA sell-out to commercial pressures (Oulton and Narayan 1995). In a neat link back to the countercuisine, among those protesting about Freedom Food is Ruth Harrison, writer of the 1960s factory-farming shocker *Animal Machines*, widely credited with setting the agenda for contemporary countercultural food ethics in Britain (Spencer 1994).

Another key contemporary community mobilisation around food issues centres on charity. In an age where global patterns of food availability stress more acutely than ever the gaps between rich and poor, acts of charity have assumed seemingly worldwide significance, often giving those involved a profound sense of their collective agency and an affirmation of caring with a 'community' locus. Live Aid stands out as a defining moment in contemporary charity, with its clarion call chorusing 'Feed the world' (the American anthem even more inclusively declaring 'We are the world'). In a paper analysing the lyrics of these two songs and the Canadian Ethiopia-famine song 'Tears Are Not Enough', David Howes (1990) explores the American inclusive 'we' against Britain's distancing use of 'they' in 'Do They Know It's Christmas?' and Canada's ambivalent 'you and I'. Additionally, the Canadian lyric is conditional (in its terms, the world *could* be changed) whereas USA For Africa's 'We Are the World' is more emphatic and refuses dissent: the world *will* change, and *we* are going to change it. Both British and Canadian songs use some elements of separation – 'Tears Are Not Enough' demands 'Let's show them Canada still cares' while the 'they' of Band Aid's title is more immediately defining of difference. Further, 'Do They Know It's Christmas?' contains some bizarre lyrical conundrums that betray a colonial approach to Africa. For a start, over half of Ethiopia is Muslim, so Christmas is meaningless; and second, the lyric tells us 'There won't be snow in Africa this Christmastime' – as if there would be if it were not for the famine. As Howes (1990: 334) says, there is a notable 'inability to conceive or portray the world other than in British terms'; the song remains located in the UK, and has an odd mix of festive celebration and world concern, captured in the couplet 'Here's to you, raise a glass for everyone / Here's to them, underneath that burning sun'. But for all this criticism of Band Aid, Live Aid or any of the other countless charity events focused on food issues, they retain a powerful symbolism of connectedness and of community action, and give those who participate something to 'belong' to, something through which to articulate their knowledge of and concern for the world.

NEGOTIATING GLOBAL FOOD COMMUNITIES

Other communities we cannot avoid belonging to. In the UK, membership of the European Community (now the European Union) has been and continues to be an issue arousing considerable debate, giving us the neologisms 'Eurosceptic' and 'Euro-rebel' to describe MPs who consistently question Britain's place within the EU. In *Appetites and Identities* (1995), Sara Delamont discusses Europe's thinking on farming (embodied in the Common Agricultural Policy together with its twin, the Common Fisheries Policy) as a response to the memories of near-starvation experienced over much of Europe in the

Second World War, and as a reflection of the continued dominance of 'peasant' farming over much of the continent (the original members of the then EEC were Italy, Germany, the Netherlands, Belgium, Luxembourg and France). Currently maligned policies such as subsidies and set-aside, and the weird landscapes of butter mountains and wine lakes, reflect a desire to regulate farming across Europe, managing surpluses, controlling prices and warding off hunger – as well as shoring up the image of meddlesome Eurocrats who want to mess with everything that's good about British food and farming. Countless urban myths (and some truths, often exaggerated) circulate about EU directives on bananas (giving us the regulation, uniform, *straight* 'Eurobanana') and potato crisp flavourings (no more 'hedgehog flavour' crisps unless there's *real hedgehog* in there), about Mafia money-making through imaginary farm subsidies, and about mineshafts and landfill sites filled with rotting tomatoes and peaches to keep prices up. Outbursts by sections of Europe's farming and fishing communities also periodically flare up: French fishermen dumped their catches in Paris and ransacked wholesalers, trashing non-French stocks in a recent row, while a return to naval skirmishes not seen since the 'cod wars', this time involving Spanish fishermen, was witnessed on the high seas in 1995.

Part of the perceived difficulty in implementing pancontinental policies on food stems from the extreme cultural diversity contained in somewhere like Europe. As Delamont describes early on in her book, at the most basic, Europe can be divided into two broad food regions: one where 'the drink is beer, the food is cooked in animal fat, and there are ample dairy products'. In southern Europe, on the other hand, 'the drink is wine, food is cooked in olive oil, and milk and butter are not central features of the diet' (Delamont 1995: 19). As we move through scales, so we see the food map of Europe divided by national, regional and local practices and habits, too. But other divisions of food consumption patterns cross countries and span continents; some reflect processes of globalisation or of urbanisation discussed later, but others trace the movements of peoples around the world. Migrant groups, often bearers of ethnic or religious identities, commonly take their food habits with them, altering the culinary culture of host nations along the way. Indeed, the 'national dishes' of countries commonly bear the mark of successive waves of migration. Delamont takes Spain as an example, quoting from Gili (1963):

> One wonders what the people of the Iberian peninsula originally ate – for olives and garlic were brought by the Romans; and saffron, black pepper, nutmeg, lemons, cane sugar, rice and bitter oranges came with the Arab conquerors; the sweet orange was introduced through Portugal from China; while the taste of *garbanzos* (chick peas) came with Carthaginians. And it was not until the discovery of America that Spain, and through her, Europe, first enjoyed potato, tomato, pimento and chocolate.
>
> Gili 1963: 10–11; in Delamont 1995: 25

Ethnic and/or religious communities within host countries are marked as different, and attempt either to maintain that difference or to erase it, through cultural processes and practices. Maud describes in Box 4.5, for example, how some of the African-Caribbean residents of Sheffield have come together to establish a community lunch-club based around sharing Jamaican foods. The very idea of difference, of course, is, as van den Berghe (1984: 395) notes, only 'fully realized through alienation'. Cooking and eating are often of central importance, as ethnic cuisine too 'only becomes a self-conscious, subjective reality when ethnic boundaries are crossed' (ibid.). In the Introduction we discussed the negotiations of identification by Turks in Germany as viewed via kebabs, and used this to illustrate the way nation, nationality and nationalism are enacted through food; but we could equally discuss this in terms of community, since Turkish workers in Germany represent an ethnic community. The distinction is to some extent blurred, but we can see how kebabs are used to make a statement about *German* nationhood rather than Turkish in this context. Although this is an artificially neat divide, it might be useful (with caution) to discuss the foodways of 'ethnic minorities' here, and the foodways of 'host cultures' under 'nation'; thus the symbolism of Japaneseness articulated in Japan through rice farming and eating is a tangibly national (and nationalistic) food practice (Ohnuki-Tierney 1993), whereas the maintenance of clearly defined food habits by Italian-Americans represents an 'ethnic community' foodway (Raspa 1984). As van Otterloo (1987: 126) puts it, 'relationships at table [i.e. around food habits] appear to be determined more by the figuration of hosts ... and guests than that of "the established" and "the outsiders"' – so the use of the term 'host' is perhaps appropriate here (and certainly easier than van Otterloo's use of autochthonous and allochthonous). The Turkish example shows how interdependent host and guest are, although the specific adaptation of kebabs to feed German eaters justifies its discussion as an articulation of nation more than of community.

In a similar case, Susan Kalcik (1984) discusses the position of Vietnamese food in the USA – an influx of Vietnamese to Kentucky led to rumours of pets disappearing into cooking pots, while elsewhere in the USA Vietnamese restaurants flourish. As Kalcik says, these two contradictory processes show how food and ethnic/national identity are commonly related to:

> In the first a food stereotype is used as a weapon against an intruder: the formula appears to be 'strange people equals strange food.' In the second process the new group presents its food in acceptable, safe arenas where some Americans try it out and learn to like it and perhaps even learn to cook it themselves. The formula here seems to be: 'not-so strange food equals not-so-strange people' or perhaps, 'strange people but they sure can cook.'
>
> Kalcik 1984: 37

Box 4.5 MAUD

MAUD IS A RETIRED JAMAICAN WOMAN WHO LIVES ALONE. SHE IS
ACTIVELY INVOLVED IN A LOCAL LUNCH CLUB FOR THE ELDERLY
AFRO-CARIBBEAN COMMUNITY.

'The lunch club came out of the necessity for, um, the need was there for us to get our people together. The first Christmas there we got together and we thought there might be people around, Christmas is coming, some of our people widower or widows, and um, you know, that are alone and might be at home on their own this Christmas so we throw feelers out and people start going round to enquire if there were any such people, and lucky enough, funny enough, there were people there. And we had, we didn't have any facility to do the cooking for the Christmas meal, so what we did, each member prepare something at home and they collected with cars, we all collected in car, each member did something, some did beef, some did rice and peas, some did a salad and then we got there and put boards together because we didn't have a facility and we had a good time, the members of the executive and a few people. We had a lovely time and, you know, talked about reminiscing and home and things like that and, oh, and we thought, "what a great idea". And so, but that was only at Christmas and then we started thinking now the necessity is there and a few people that work in the health service, few of our members said they'll try to organise meals on wheels with the authority if we could get lunch to go round to people. We didn't have any restaurant we didn't have anywhere to cook it.

'But when the community worker went to investigate about this she, we were told that we had to go to a cold storage depot and get the meals, the frozen meals like what the English people are doing, you know the, um, and take it and you take it home and heat it up. And we said WHAT! no way, because we don't want that sort of food what they prepare. So we said no, no, no, we want, that wouldn't be good enough for our people [the Afro-Caribbean community], we want our people to have the food that they're accustomed to having, you know. We want our rice and peas and we want our green bananas. Want our dumplings, and we want our yams. We want the food we want. So that's how the luncheon club was born. So as I say, we were in the process of moving down to this centre and when we got down there we decided to prepare the meals ourself so that's how the lunch club was born.

'The Jamaican diet is healthier than most diets but it tend, there again, ahm, a lot of starch and that's not – you see Jamaicans have, a lot of Jamaicans have got diabetes because they, they concentrate on the high concentration of starch in the diet, you see. And sugar, we love sugar, because one of our main – we, we grate the carrots and juice it and then you mix it with condensed milk and maybe put some Guinness in there sometime, but with the condensed milk you got the natural taste and that's one of the main things. It's lovely, but it's sweet, yes, it's all sweet and the condensed milk and thing like that it.'

This is a very interesting way of approaching the food–identity nexus; in Kalcik's analysis, the Vietnamese food infrastructure evolved to serve Vietnamese customers, providing food as a way of maintaining and celebrating ethnic identity. Then two processes of 'acculturation and hybridization' (Kalcik 1984: 39) take place: the ethnic cuisine becomes modified to suit local food habits (what ingredients are more readily available) and gradually incorporates more elements of the 'host cuisine' (in some cases the 'host' state facilitates or even enforces this, often in the guise of 'home economics' tutoring for low-income ethnic households and so on); at the same time – and in a process often linked to this 'watering down' of ethnic foods – members of the 'host' community begin to sample the ethnic cuisine, to get used to its presence, and frequently come to enjoy it. By this time the food itself may be far removed from its original form and meaning, as the case of doner kebabs in Germany illustrates. Choosing to adapt traditional culture in the context of new national and local codes, and in the wake of transnational and global changes, calls for what Bhachu (1995) calls 'cultural entrepreneurship' in which culture (including cuisine) is continually remoulded – as Kalcik (1984: 56) succinctly puts it, 'one does not have to be authentic to be ethnic'.

Kalcik thus emphasises the need to think of ethnic identity as processual and performative rather than fixed. Within this, the role of 'ethnic food' is of great importance; its dynamism – reacting to 'host' culture (and *being reacted to* by 'host' culture) – shows the malleability of foodways in the negotiation of identifications. Eventually, of course, the traces can become all but lost, or incorporated into a hybrid culinary culture which over time comes to be seen as 'traditional', as Gili's archaeology of Iberian cuisine showed. In other contexts, of course, these traces are not lost but reiterated constantly, in the maintenance (or even reinvention) of a separate, definable *ethnic* identity. Thus Raspa's (1984) study of the Italian-American community in Mormon Utah found the symbolic importance of keeping a distinct Italian cuisine for community members was central to what he called a *nostalgic enactment* of identity. The consumption of foods viewed as traditional by 'insiders' and as at best unappetising by 'outsiders' – such as goat meat and especially goats' heads – is a powerful statement of identity and difference, but also a nostalgic and 'invented' one.

As was stated earlier, it is not just recipes and dishes which need to be considered when discussing ethnic cuisine; other elements of food habit are also retained. A study by Goode, Curtis and Theophano (1984) of meal formats, cycles and menu negotiations in Italian-American households shows continuity in the *infrastructure* of meals, while Tejano (Mexican migrant workers in Texas) women's strictly maintained habit of laborious food preparation for husbands is seen by Williams (1984) as a 'revisionist' tendency for affirming traditional family life. Likewise, cooking techniques and equipment carry important symbolic weight in ethnic cuisines.

Kalcik's (1984) discussion of the performance of ethnic identity is important here because it stresses the importance of the site of performance and its audience. Whether attempting to maintain difference, show willingness to become 'American' (for which read WASP), or accept pluralism and hybridity (these three tactics Kalcik lists as the three communicative messages of the performance of ethnic identity), there is a clear distinction to be made between private and public performance, and in-group or out-group audience. The contrasting Italian-American home and restaurant food habits illustrate this, with the latter catering for non-Italian tastes and meal formats. As an expression of ethnic pride, on the other hand, the public provision of 'authentic' ethnic cuisines for 'outsiders' – at festivals, for example – might contrast with an Americanised diet eaten at home.

The role of ethnic restaurants in culinary-cultural change cannot be overstressed. General trends towards eating exotic regional and ethnic foods have been widely remarked upon by commentators. The career of ethnic food business mentioned earlier, which begins with provisioning in-group consumers and later branches out to provide for 'host' culture (over time, effectively becoming 'naturalised') has been charted time and again for all manners of ethnic food (e.g. De Vita 1994; Kraut 1979) and is widely seen as a vital component in the emerging foodscape of any country. Zelinsky's (1985) 'roving palate' paper maps ethnic restaurants across the USA, revealing the character of both 'hosts' and 'guests' as they meet at table. And a case study of Mexican restaurants in Tucson by Arreola (1983) shows how an increased demand for Mexican food has led not only to a proliferation of outlets but to their increasingly trading on ethnic identity: a new Mexican restaurant architecture reflects a growing 'ethnic consciousness' which makes sound business sense, too. As Arreloa (1983: 112) concludes: '[t]he exotic cuisines, it seems, requires an exotic exterior to impart authenticity' – heightening the contradiction by being a new, invented cultural landscape form, to match what is in fact a hybrid cuisine often prepared and sold by corporations completely disconnected from anything 'authentic' or 'ethnic' (Belasco 1989; Paulson-Box and Williamson 1990).

The taste for exotic and ethnic foods reflects a certain kind of attitude – one which is often referred to as 'cosmopolitanism'. A true cosmopolitan would, of course, pour scorn on corporate chains peddling 'inauthentic' cuisines, and search for out-of-the-way places where the genuine article is still available, untouched by homogenising forces. In the next chapter we connect the theme of cosmopolitanism with its dominant scale of expression; for if the cosmopolitan has a home, it is the city.

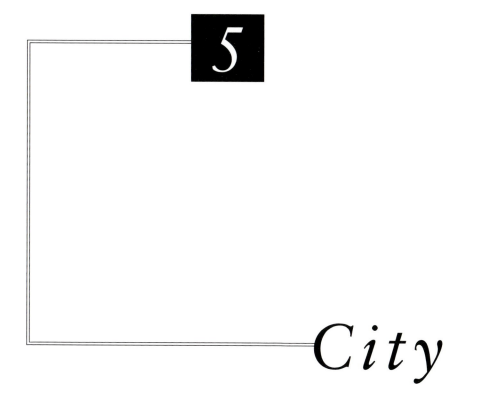

5

City

YI-FU TUAN (15)

●

The fact is I don't and can't cook. When I have to cook, I do the following, which ends in a sort of decently nutritious noodle soup. One cup of water, half a cup of Campbell's chicken broth to be placed in a pot, and just when the mixture is about to boil, add six dried shrimps, four black (dried) mushrooms, a package of noodles (the sort that cooks in five minutes), half the content of the flavour capsule that the package includes, and a handful of frozen Mediterranean vegetables. Boil the whole mess for another five minutes and serve. For accompanying beverage, I recommend Ginseng tea, grown in Wisconsin, packaged in China under the directions of headquarters which is located in San Francisco.

CITY

Urban living is a rich and complex business. So is urban dining, since the city has witnessed an explosion in eating places (and shopping places) which can be bewildering or bedazzling, depending on your perspective. Walking along the high street, or through the aisles of a supermarket, is a truly mouth-watering experience. Restaurants, fast-food outlets and food shops are all around us, offering up every imaginable (and unimaginable) culinary temptation.

In this chapter we look at restaurants, both as places to eat and places to work, examining their key role in the renaissance of cities as sites of cultural capital. We explore many forms of urban eating beyond the formal restaurant, however, reflecting on the built landscapes of urban food consumption and on the sociocultural significance of their ever-evolving forms. We then move on to consider that other temple of urban consumption, the supermarket. The urban morphological impact of out-of-town supermarkets has been incredible, as has their effect on the daily (not to mention weekly, monthly and yearly) rituals of city living: the trip to the supermarket is now a landmark in the structuring of our life-patterns. Such is the importance of food to cities, in fact, that urban promotion is increasingly seeking to capitalise on a city's culinary associations in the race for prominence on the world-city stage.

5

CITY

The restaurant has unwittingly become a symbol of contemporary [urban] life. Our love affair with the automobile, the flight to the suburbs, the return to the city, disposable materialism, and even the counterculture – all are reflected in this constantly changing milieu of Alice's Restaurant, McDonald's, and Clarisse's Soul Vegetarian Cafe. Dinks, yuppies, an increased disposable income, moving away from the mother-in-law, car pools, Little League, the long trek to the exurbs, hating to empty the dishwasher, and that continual societal evil, television – have made their contributions to the restaurant revolution as well. . . The restaurant . . . clearly has become a mirror of ourselves, our culture, and our new geography.

Pillsbury 1990: 10–11

Richard Pillsbury's critical review of the place of the restaurant in contemporary American culture, *From Boarding House to Bistro* (1990), has an unmistakably traditionalist tone; his charting of the rise of mass consumption and what he calls 'disposable materialism' is always expressing the uneasy displeasure of someone who eats Big Macs but knows he really shouldn't. His emphasis on eating out as a cultural barometer, however, is particularly interesting in our consideration of the scale of the urban as a site of food consumption; as Pillsbury himself acknowledges, there are important dining-out sites away from the city – secluded country restaurants, village tearooms and so on – but for sheer volume and choice, the city (and especially the large city) remains the prime location for that defining activity of cultural distinction: eating in the public sphere. In *Ronald Revisited: the world of Ronald McDonald* Marshall Fishwick (1983: 5), wrote an ironic, Beat-ish road poem commenting on the proliferation of eating places along the highway:

Shakey's Pizza, Tastee Freez,
A & W, Hardees.

Howard Johnson, Red Barn, Blimpie,
House of Pizza, Big Boy, Wimpy.

Wendy's Friendly's, Taco Titos,
Sandy's, Arby's, Los Burritos.

Surveys of the whole eating-out business, of specific sectors of it (e.g. ethnic restaurants) and of individual chains pile up evidence of an incredibly important social (and geographical) phenomenon – for Fishwick's poetry maps the flickering of fast-food outlets along the streetscape, as his car cruises by (see Belasco 1989; Bloomfield 1994; Pillsbury 1990; Zelinsky 1985). In this chapter we take a similar cruise through the city streets, noting those urban landscape forms built around food: the restaurant, the take-away, the supermarket and so on. We can't promise anything poetic, but we will try to trace some links between the city, food and eating, and identity. We start our trip at a common sight next to British city roads: a 'family restaurant', which is an interesting place to begin pondering eating out as a social activity.

EATING OUT: THE CONSUMPTION PACKAGE

In a recent newspaper article, food writer John Lanchester visited one link in a British eatery chain – Harvester restaurants – and commented on the chain's approach to making its customers lose their unease at eating out, hence enlarging Harvesters' mass appeal as 'family restaurants' (rather than just glorified purveyors of 'pub grub'):

> Harvester restaurants make you feel the extent of restaurant-fear through the intensity of their denial of it. The emphasis is so assertively unpretentious that it's almost hysterical; it is about fear, in the same way that positive thinking self-help books are about failure.
>
> Lanchester 1995a: 45

Despite such attempts at 'democratising' the eating-out experience, it remains a heavily stratified activity; in fact, this is its selling point, whether in terms of Harvesters' hysterical openness or the intense food snobbery of more 'high-class' joints. And, of course, this is all because of its location: in public. No one need know what we eat at home, behind closed doors. But in a restaurant, everything we eat and the ways we eat it are on constant display, under continual surveillance. A carefully managed environment must thus create the suitable atmosphere for food consumption before an audience of fellow diners (Finkelstein 1989). As Beardsworth and Keil (1990: 142–3) note,

> [T]he restaurant exists as a feature of the entertainment industry, and is as much concerned with the marketing of emotional moods and desires as with the selling of food. . . . Eating in the public domain becomes a mode of demonstrating one's standing and one's distinction by associating oneself with the ready-made ambiance of the restaurant itself.

The eating place, then, offers a total consumption package – not just food and drink but a whole 'experience'. The Harvester chain produces an environment as fear- and guilt-free as possible, with an unfussy menu, specials for kids, and an emphasis on good value. People who might otherwise see eating out as an upper-class luxury are thus reassured that they are not out of place here. But where has this immensely popular (and meaningful) social activity come from, and when?

In an interesting note on eating out's history, Mennell, Murcott and van Otterloo (1992) remind us of the availability of 'street foods' at least as far back as the medieval period – countering the widely held belief that restaurants were an unintended byproduct of the French Revolution, which had created a body of unemployed cooks who had previously worked for the aristocracy. Precursors of take-aways, fast-food outlets and *haute cuisine* restaurants litter historical records, from inns to coffee-houses, from cookshops to cafes and taverns (the latter being the closest antecedents of today's restaurants). Over time, social changes combined to produce a body of informed and discerning consumers, reshaping both the cultural forms and the market relations of eating out (see also Mennell 1985). And as Habermas (1962) noted, venues for eating out were important in the development of the public sphere in the modern sense, since gossip, commercial intelligence and political intrigue were always on the menu, too – so much so that, through most of the nineteenth century, the Parisian police force used cafe spies to infiltrate 'dangerous political elements' (Sennett 1994: 347). As well as being a centre for information-gathering and politicking, the cafe operated as a labour exchange, a university, and a dating agency: '[T]he cafe was also a meeting place for lovers, an erotic as well as a business setting' (Wilson 1996: 89). As Wilson notes, today's cybercafes, with their Internet link-ups, may put some of these functions back into cafe culture.

The act of eating out – whether the intensely scripted interactions of an *haute cuisine* restaurant or the free-and-easy casualness of pub grub – is, then, a container of many social and cultural practices, norms and codes. Both those who eat and those who serve are subject to these, enacting a stylised set of encounters famously described in Erving Goffman's 'dramaturgical' account, *The Presentation of Self in Everyday Life* (1959). Although his work, and much of that which has followed, focused on those preparing and serving food (Crang 1994; G. Fine 1988; B. Fine 1995; Mars and Nicod 1984), the behaviour of diners is no less mannered and structured (Finkelstein 1989). Everyone involved is part of an elaborate 'place ballet': Roy Wood (1995: 103) goes as far as to call restaurants 'theatres of the absurd'. Anyone interested in contemporary consumption practices is now well aware that such structuring permeates all 'interactive service encounters', whether in a shop, on a plane, or at any point of sale (Du Gay 1996). Phil Crang (1994: 677) calls these 'sociospatial relations of consumption', reflecting his interest in the *geographies* of such encounters.

One of the most remarked-upon outcomes of this emphasis on the performative nature of food service work is the policy of selecting and recruiting waiting staff on the basis of their personalities – or, more accurately, on their self-presentation of certain attributes of their performative personalities (extroverted, 'fun-loving', sociable, chatty, etc.). Artists and 'resting' actors are thus often sought (Zukin 1995), as are suitably camp gay men able to provide 'a bit of homosexual frisson' (Crang 1994: 691). Crang's account is wonderful at providing a detailed picture of how this increasingly common aspect of service work (made popular by management gurus' obsessions with Disney-world workers) turns waiting staff into 'part of the cultural capital being sold' by restaurants (Crang 1994: 693). Work by Sharon Zukin and research students on New York eateries focused particularly on the employment of 'artists' (including actors and musicians) and their contribution to the 'symbolic economy' of the restaurant. As she writes,

> [A] widespread perception of artists' importance to a restaurant's ambiance has helped renew the reputation of restaurants as centers of urban cultural consciousness.... Waiters are less important than chefs in creating restaurant food. They are no less significant, however, in creating the experience of dining out. For many people, oblivious of restaurant workers' social background, waiters are actors in the daily drama of urban culture.
>
> Zukin 1995: 154

Those daily dramas of urban culture hinted at by Zukin can take many forms, and the restaurant can accommodate most of them. A newspaper article by Henry Porter (1996) describes precisely this: 'We don't just dine out in the Nineties – we live out, marking births, deaths, marriages and every possible partnership, concord, disagreement, alliance and dissolution, all within the earshot of at least half a dozen strangers' (Porter 1996: 32). While gossip, political agitating and love-matching were common social functions of the bohemian Parisian cafe culture, the contemporary restaurant is also a site of business deals, 'gladhandings' (conspicuously public greetings), firings, acts of revenge, even killings. Such 'hostile uses of the lunch appointment', as Porter calls them, are part of the 'modern carnival' that is eating out. As one (anonymous) restaurateur put it to Porter, the secret of a good restaurant is

> not [to] let the food submerge the rest of the culture. Too often restaurants try to dominate people with the food or the chic intimidation of the waiters. The good restaurant acts as a catalyst for people to do what they want.
>
> Porter 1996: 34

And what they want to do is clearly much more than just eat.

Porter's quip about the 'chic intimidation of the waiters' is echoed in recent reports of surly table staff in London's elite eateries (Moir 1996). While the rude waiter is a standard comedy figure (a US 'fanzine' for restaurant workers, *Eat ... and Get Out!* had its first editorial on the theme of 'service with barely concealed contempt' (Glenn 1995)), an interesting feature remarked upon by Moir is the willingness of British patrons to accept such treatment without complaint, laughing it off as part of the meal experience. In a world of nauseatingly 'nice' interactive service encounters, perhaps a bit of aloof surliness comes as welcome relief? Or perhaps it is an inevitable strategy of resistance for those cajoled into wearing a happy face and being perennially hyper-polite (Holliday, forthcoming)? As Crang (1994) shows, the pressures of being constantly on show while doing a basically mundane set of tasks make any acts of symbolic refusal important for workers' self-esteem and self-identity.

In spite of this outbreak of surly waiters, eating out is an increasingly popular leisure activity in Britain, remarked upon in many social barometers, including *Vogue*, which has declared food 'the new sex', and quotes one Soho restaurateur as stating that 'Young people are bored with clubbing and restaurants are becoming more of a social thing' (Foulkes 1996: 43; Wood (1992) makes a more general point about the under-researched connections between eating-places and youth culture). Trendy eating-places are springing up all over (though with an unsurprising concentration in the capital). Terence Conran's massive Mezzo, with 700 tables, the phenomenal success of noodle bar Wagamama, and even the proliferation of Seattle-inspired coffee-houses are all taken as signs of a renaissance in dining out. One critic suggests that this revival is about escaping 'mundane English reality':

> In other cultures, to eat in a good restaurant ... is to commune with the soul of one's nation or region or city, to enter into a sort of dialogue about identity. In London, it is to flee – whether to Thailand, to Provence or, at Mezzo, to a fantasy of the Jazz Age.
>
> Popham 1995: 2

This desire to escape – the 'urge to party in strange spaces' as Popham (1995: 3) calls it – has led to an incredible diversity of eating places in London, with some quite dizzying concentrations of difference: the Uxbridge Road, for instance, is described by Matthew Fort (1996b: 25) as 'one of the most polyculinary streets in London', whose many restaurant menus 'resemble travel brochures'; Will Self punningly calls it a 'trougherfare', or a 'great artery of a road, lined with pulsing eateries', with 'hundreds of establishments from the East lining the route west' (Self 1995b: 57). Soho, once renowned for its sex industry, is becoming '*the* restaurant district of London' (Foulkes 1996: 43), with porn cinemas being converted into chic eating-places. And Ian Cook and Phil Crang (1996) recount the description of London's eateries as offering 'the world on a plate' in the listings magazine *Time Out*, which in 1995 implored readers to 'Give your tongue a holiday and treat yourself to the best meals in the world – all without setting foot outside our fair capital' (quoted in Cook and Crang 1996: 131). Some of the pleasures of eating out – especially in new places – are captured in Anna's account of dining out with her parents in London (Box 5.1). But not everyone has the licence to experiment in this way; for those restricted by income or cultural traditions, the city foodscape does not offer the same exciting new possibilities, as Rehana explains in Box 5.2.

Box 5.1 ANNA

ANNA IS IN HER LATE TWENTIES AND WORKS FOR AN ENVIRONMENTAL CHARITY IN A DERBYSHIRE VILLAGE. SHE GREW UP IN LONDON, WHERE HER PARENTS STILL LIVE.

'My parents are very keen on eating out, probably mostly to do with the fact that neither of them can stand cooking. I think it was early teenage years they started to take my sister and I along with them. Before that, if they ate out, they'd probably leave us at home and granny would look after us. They tended to have two or three restaurants they went to very regularly – very little adventurous spirit there. There was one called Changing Times they used to go to about every two weeks, run by Ted Knight who used to be leader of – well, deputy leader of the GLC – Red Ted [laughs]. [Ted Knight was in fact leader of Lambeth Council in London, not deputy leader of the Greater London Council.] That was a wonderful place – very odd and very eccentric staff.

'It was sort of slightly English, French, Scottish – somewhere in between. It's very difficult to describe – it tended to be quite sort of meat-based and piles and piles of various sorts of vegetables in interesting sauces, and they used to use huge platefuls, but you'd wait several hours to actually get anything, and then the cook would have a tantrum. It was just round the corner from us and they were always so relaxed and laid back that we used to go there very, very often.

Box 5.1 Continued

'And they were very well known there. They like going places where they get to know the staff, where they tend to know the menu off by heart. Now they always go to an Italian just round the corner – that's a cross between Italian and French – and again, they know the person who owns it very well and the staff, and they always take the mickey out the staff no end, 'cos they tend to hire Italian girls from a local language school who have just come over for a year to learn English.

'That place is – both those places, they do seem to be very much regular-type haunts, 'cos it's in a suburb of London and I think people eat – well, people who want to try something different or something new, tend to go up town or something like that, whereas if you stay five minutes away from your door it tends to be very much the regulars who go time and time again. Plus also, they like eating out mainly because it's just relaxed and not a chore, so they tend to just like to pop round the corner to the local restaurant where they feel at home and the food's good.

'Whenever I go down there or my sister visits, they'll tend to make a point of going out as a sort of treat, and then they actually sometimes do go somewhere different. I think a lot of it's to do with the fact that I like the more exotic food – I actually prefer things like Indian and Chinese to standard British fayre. Plus also, the place they live in has quite a large sort of fashionable-type community. It's on the edge of Clapham Junction and Battersea where you get all the sort of yuppie types, so there's a lot of restaurants down there which are sort of ethnic and a few nice, green-type veggie restaurants.

'My favourite one of all is just off Leicester Square – it's a vegetarian southern Indian restaurant which is incredible – it's the best Indian food I've ever tasted. Giovanni's is a very nice Italian, but it's quite dull, ordinary Italian, whereas this Indian one is sort of quite special as far as I'm concerned.

'Nearer home, I tend to like to go somewhere I haven't been before, so usually something a bit unusual – there's a couple of South American restaurants recently opened down there. There's a reasonably large Columbian population in that area of London – mainly political refugees of one sort or another. But that's the wonderful thing about that part of London – there's so many different peoples actually come in and set up home there, so the food shows that. Plus as well, you get this sort of yuppie set who like the ethnic food as well, so it's not in a way, sort of, for their own culture and people, it's more to, a trade to bring in other people. Having said that, the one place I tend not to have much success in dragging them is over to Balham, where there's some wonderful Indian restaurants which are properly for Indians [laughs], and that is where the food is a lot better than anywhere else. But that means venturing out.'

Box 5.2 REHANA

REHANA IS 28 AND HAS FOUR CHILDREN. HER HUSBAND IS A TAXI DRIVER IN SHEFFIELD.

'We [Muslim women] can't eat without our husbands in Asian restaurants in Sheffield, we could eat out of town but not in town 'cos, well, you know how you are in your Western community, you just live your life, don't you, you just do your own thing. But our community, they just drag you down, everyone talks about you behind your back, there's a lot of stabbing in the back, it's sick, it's just one of the sick things about our culture, you just – and if you went in an Asian restaurant in Sheffield, oh, there'd be, it's not worth it. We went to, my friend's husband had a restaurant actually, in Sheffield and we went there once for her birthday, went for a meal and that, and I had to, oh, I had to be "the Muslim woman", I had to be so good. And I'm not like that, I like to lark it, you know, have a laugh, but I had to be a real woman.

'Because I drive his car, like, in the day when he's not working, and his friends they all flash at me because they see the car and they think it's him. So I can't do anything, they know where I am because they recognise his car and know it's his wife. And then he goes to work at about six, then after work he goes to an Asian restaurant for a meal, or gets a take-away. Because they all eat there, you see, in the Asian restaurants, they all go there after work at two and three in the morning, and they all talk and talk. I would never set foot in that place where he goes, he's there every night, they all talk there. It's weird, isn't it, while we're asleep they're having a great big dinner after work. But they talk, it's "your missis did this" and "that woman did that", and you can't, you can't just go out because – he wouldn't let me eat in an Asian restaurant in Sheffield because they'd be talking, he'd say something about someone doing such and such, and someone would say "well, we saw your wife out" and he'd be shamed, it would shame on him. He'll say things about other people's wives but he doesn't want anyone to talk about me. You can't just get on with your own life, like. So I could go somewhere like Meadowhall, 'cos that's like just a big place, it's not specially going out for a meal, it's just lots of people together. But I couldn't go in the centre of town, everyone knows everyone and it wouldn't be worth the talk. I'd have to go to Bradford or Leicester or somewhere, and then I'd have to get the car.'

URBAN FOOD RITUALS

But it is not just in concentrated 'eating districts' of the city that dining out is flourishing. A companion phenomenon remarked upon by Roy Wood (1995) is the penetration of food services into almost every other leisure site: theatres and art galleries, leisure centres and cinema multiplexes, shopping malls and sports stadia, supermarkets and airports are increasingly places to 'catch a bite' while also catching a play or a plane (Lowe and Wrigley (1996: 27) note that 'filling station snacking . . . is the fastest growing but least predicted area in retailing today'). As Wood (rather disapprovingly) writes:

> [D]ining out is no longer the singular activity it once was; it is not always an activity pursued for itself or in itself, but associated with other, mainly leisure, activities which are all part of a loose-knit collection of consumption-oriented services dedicated to making people enjoy themselves.
>
> Wood 1995: 111

Almost everywhere we go, then, there are opportunities to eat. The presence of food in public space can, of course, still be unsettling, such are the rigidities of the codes of conduct which govern it, even in the midst of its overabundant availability. Eating outdoors, especially on the street, remains taboo for lots of people – a sign of lacking social graces (except in culturally sanctioned contexts such as picnics; see Hartley 1992). It is also an intensely gendered form of consumption in many societies where women's presence in public space is closely regulated (e.g. Cowan 1991). In Britain, Victorian suburban fears of the menace of roving food vendors produced a legislature on street trading which in part remains, and which has contributed to popular conceptions of the moral geography of polite public food consumption (Burnett 1983).

Even within approved locations, food's presence can still be ambiguous. Most Western diners are forbidden from consuming food not bought on the premises when they sit at a restaurant table (but here again there are exceptions, such as feeding babies with ready-prepared mush). Of course, this is primarily a commercial ruling, in the same way that a lot of eateries frown on non-customers using their toilets, and will not let you sit at a table unless you are going to eat, nor permit you to linger needlessly after you have finished (Helmer (1992: 95) refers to the informal restaurateur's motto, 'Eat it and beat it'). But there are also countless other conventions about eating out. One doesn't, for example, eat the previous customers' leftovers. Eating from each other's plates is often seen as impolite (except in certain cuisines where plates and dishes are used more communally), as is starting your main course before your fellow diners have finished their starters. Eating too quickly, too slowly, eating too much or too little, what you select from the menu (especially if someone else is paying) . . . every move you make is governed by some sort of restaurant norm, making eating out an intensely structured affair.

Anything which does not fit this structure can therefore be quite disruptive. A conference paper by Michael Maynard (1995), for example, considered some bananas which had been abandoned or left on the table at a shopping mall's food court – a bunch of perfectly eatable bananas, out of place in the midst of conspicuous food consumption. Maynard observed customers avoiding the bananas, even moving them to another table, so that the table they had previously lain upon could be used to sit at and eat. Starting with his 'unclaimed bananas', the author embarks on a complex decoding somewhat reminiscent of those undertaken by Nicholson Baker's narrator in the novel-of-the-lunchbreak, *The Mezzanine* (1986). For where Baker's hero considers in some depth – almost obsessively, in fact – the contents of his sandwich (as well as everything else from shoelace tips to escalators), Maynard asks, 'Is taking a couple of unclaimed bananas taboo?' (2). Even though they were unclaimed, going free, no one ate them, or took them home (not even Maynard, who confesses that bananas are his favourite fruit).

The bananas in Maynard's analysis are 'devalued food'; in an environment of abundance, of wealth, and of the celebration of the human production and consumption of foodstuffs, unclaimed bananas are truly out of place. But it is in his final reading that Maynard hits on an especially powerful subtext – that of fear. Unclaimed bananas just aren't trustworthy. Their biographies are not exposed like those of the foods prepared before our eyes, for our consumption. They might be tainted, or adulterated, or in some way or another uneatable. As Maynard writes, 'the unsponsored bananas seem violently out of place because they are without "papers." Nobody (no human) is there to vouch for them; nobody (no human) to claim responsibility for their authenticity or goodness' (5). They become forbidden fruit. Worse still, they might have been placed on the table by an academic wanting to note how customers react! Their ultimate fate? Discarded by food court staff in a routine cleaning sweep of the tables, still unclaimed, still uneaten.

What Maynard's thick description of the brief career of these unclaimed bananas does is help us see just how highly structured the eating-out environment is, and how complex its rituals are. Perfectly good, *free* food is viewed with suspicion, and ultimately thrown away. Fear of adulteration haunts many who eat out, and tales of domestic pets, cockroaches and even the occasional human body-part ending up on a restaurant plate litter urban folklore. More common than that are suspicions of substandard ingredients and overpriced meals: restaurants rebottling cheap plonk to sell as fine wine, or using meat destined for petfood in place of sirloin steak. Add to this complaints about the standard of the cooking, quarrels over service-charges and countless other points of conflict, and it seems surprising that anyone chooses to eat out. But, in fact, more and more people are doing so. Chains like Harvester are popularising the restaurant, something which fast-food outlets are now realising: the hamburger chain Wimpy recently ran a TV commercial stressing that eating there is a 'proper' restaurant-like occasion, with crockery and cutlery, and service at your table rather than the minimalist in–out production line more typical of fast-food joints.

FOODIE CITYSCAPES

In his excellent survey of American chain restaurant architecture, Philip Langdon (1986) observes shifts even in eatery design as responses to the changing cultural conditions of the USA. In the 1970s, for example, a number of chains adopted a style which blended natural materials (especially wood) with jaunty, angular architecture, thus suggesting 'rustic functionalism with a sheen of sophistication' (172). Previous decades saw a similar tally between restaurant architecture and the nation's social tone: in the 1950s, plastic, neon and futuristic designs reflected the nationalistic, technophilic optimism of the space race, an optimism lost in the urban troubles of the 1960s, during which time fast-food outlets mellowed and turned nostalgic, becoming the faux-urban monuments of the 1970s. As Langdon (1986: 165) says:

> [C]hain restaurants were superbly prepared to embody popular expectations.... [They] were free to act as a barometer of public mood.... By the middle of the 1970s, restaurants recorded an entirely different mood from that of twenty years before. Americans felt more in need of solace than of stimulation. The country was dispirited, and it wore its feelings on the roadside.

In the 1980s, the playfulness and referentiality of postmodern architecture infected even burger bars and coffee-and-doughnut drive-thrus (Bloomfield 1994), harking back to the free-for-all 1950s ostentation, but adding some postmodern irony. This trend is reflected elsewhere in culinary culture, too. Advertising is perhaps the most self-conscious appropriator of cultural trends, and an advert republished in John Hartley's *The Politics of Pictures* (1992: 58–9) contrasts cubism with postmodernism as representations of instant gravy. Chefs and cookery writers have likewise turned to the playfulness of postmodernity, throwing odd juxtapositions of ingredients and culinary styles together in restaurants of equal polymorphousness. Here is Elizabeth Miles, writing about one renowned pioneer of Californian cuisine:

> Wolfgang Puck creates cuisine that both expresses his own identity(ies), and mirrors what he sees as the identity(ies) of his customers ... these identities reflect the multi-ethnic, multi-cultural, multi-gendered, nomadic paradigms of post-modernism ... Puck brings this nomadic sense to the stovetop when he cooks.
>
> Miles 1993: 193

Most famous for the 'California pizza' and for 'Asian–French' cuisine, Wolfgang Puck serves dishes pastiching culinary traditions and bringing together unexpected elements – such as the warm sweet curried oysters with cucumber sauce and salmon pearls, followed by *mille-feuille* of squab with wild mushroom sauce ordered by Miles on a visit to Puck's celebrated Chinois restaurant in Santa Monica – a restaurant deploying familiar postmodern motifs in its architecture of fake high-tech mixed with Chinese kitsch (designed by Puck's wife, Barbara Lazaroff); the referentiality even extends to the waiting staff, who are all Caucasian but wear simulacra of East Asian dress. Back home, you can try your own hand at replicating – or, even better, parodying – these masterpieces by referring to one of Puck's cookbooks. Packed with obscure and exotic ingredients, his recipes 'presume a great, diverse natural bounty magically melting from specific farmland and oceanic locations into the decentered city. This is the postmodern landscape, where nature meets city in a seamless continuum of goods and commodities' (Miles 1993: 199). We shall return to this playful, postmodern flirtation with the exotic in Chapter 8, with a discussion of the 'glocalisation' of food consumption; no doubt we'll return to Wolfgang Puck there, too.

In many other ways, eating places seize on prevailing social trends and turn them into slick marketing packages, as Helmer's article (1992) on McDonald's advertising in the 1980s shows. Reacting on the one hand to stiff competition from rival chains, and on the other to America's renegotiation of the meanings of home, family and community, McDonald's set about marketing love and togetherness through 'slice-of-life' vignettes (comparable to several UK TV ads, including the couple out shopping – woman indecisive about new clothes, man unable to choose from McDonald's vast menu – and the famous 'reunited family' advert where a boy engineers his separated parents to bump into one another at their local McDonald's). The whole concept of the 'Happy Meal' signifies that McDonald's offers much more than just food. As Helmer (1992: 94) puts it, 'McDonald's elevated the issue to a plane above simply satisfying hunger: Everyone will get enough to eat. The question is, who can give you a sense of place, where is home, who loves you?' The answer, of course, is staring up at you from its styrofoam packaging.

McDonald's has, as readers may by now be aware, been the focus of an immense amount of academic scrutiny, most of it critical. It is important to note here the significance of fast-food provision in the urban context as something which has had a profound effect on city life. Roy Wood (1995: 110) describes 'dining out to dine in' – the take-away – as a crucial feature of contemporary city cultures, contributing to what he calls the 'interpenetration of public and private cuisine'. The flourishing home delivery business means that the 'pizza lifestyle' is available to almost everyone (Oxford 1995: 2), offering the ultimate in convenience dining. And as supply gallops to keep apace with demand, fast-food outlets have become the most common first job destination in the UK (Lash and Urry 1994). Incidents such as the murder on-site of three co-workers by a Washington McDonald's employee, the fuss over 'super-sizing' (the trend towards impossibly large portions such as the McDouble and the MegaMac), and the bad-publicity vehicle that is the 'McLibel' trial might reveal something rotten at the core of the fast-food business, but the expansion figures suggest insatiable appetites for its fare (McDonald's has over 18,000 outlets in ninety countries; Pizza Hut has 330 branches in the UK alone).

It is homogenization which is most commonly attacked as the negative cultural contribution of fast-food outlets – the line between dependability and monotony being a fine one. A particularly nuanced critical account of McDonald's is provided in Susan Leigh Star's (1991) discussion of being allergic to onions. From the perspective of technological standardisation and mythical flexibility, Star's own allergy to onions lets her critique the ordering rationalities that govern how McDonald's functions:

> McDonald's appears to be an ordinary, universal, ubiquitous restaurant chain. Unless you are: vegetarian, on a salt-free diet, keep kosher, eat organic foods, have diverticulosis (where the sesame seeds on the buns may be dangerous for your digestion), are housebound, too poor to eat out at all – or allergic to onions.
>
> Star 1991: 37

Star moves from this consideration to talk about living 'with the fact of McDonald's' even if your actual participation in it as a customer is curtailed (for instance by the fact that you can't eat onions) and, importantly, to think (using Donna Haraway's notion of the cyborg as the relationship between standardised technologies and local experience) about the place of the non-user in critiquing the seeming inevitability of McDonald's – to think that 'it might have been otherwise' (Star 1991: 38). It is a complex set of arguments, but one which is immensely suggestive of creative ways of approaching theoretical questions through something as mundane as the onions in a hamburger.

The McDonaldising forces of fast-food homogenisation are, of course, only one manifestation of contemporary trends in eating out in the city. And in urban cultures, Puckish cosmopolitanism is a constant antidote to the same-everywhere dependability of burger chains – an exotic 'other' to set against the familiarity of home. Ulf Hannerz (1990) has produced a description of the cosmopolitan as someone who both masters and surrenders to 'alien culture' (so as not to be confused with, for example, the hated 'tourist'). As Hannerz (1990: 240) says, cosmopolitans will go to some lengths in order to immerse themselves in authentic 'other' cultures: 'Some would eat cockroaches to prove the point, others need only eat escargots.' As Jon May (1993) argued, cosmopolitanism is often associated with 'new service class' members (a class segment increasingly chased by marketers (not to mention academics), picked out in market research reports as trendsetters and as heavy buyers of 'ethnic' and 'exotic' foods; see *AgExporter* 1989; Miller 1995). It involves the cultivating of 'globalised cultural capital' as a form of lifestyle shopping which, crucially, involves possessing considerable knowledge about 'the exotic', 'the authentic', and so on – often referred to as a colonisation or an intellectualisation of popular culture. Food media increasingly provide such knowledge (called by Phil Crang (1995: 5) 'edible cultural geographies') to an ever-hungry audience, while urban food shops (both proliferating specialty shops and sophisticating supermarkets) search out the ingredients to help us make 'real' global cuisine – a cuisine also readily available in countless restaurants, cafes and diners. Chapter 8 takes up this theme in much more detail, but we will stay with food shopping for a little while here.

SUPERMARKET SOAP OPERAS

It is often remarked that work on sites of consumption has focused on the spectacular at the expense of the commonplace (Jackson and Thrift 1995). Thus, fantastic shopping malls are often visited in academic texts, but corner shops usually get overlooked. However, the routine acts of the weekly grocery shop or nipping out for a pint of milk deserve close attention, for they are no less important urban cultural activities than *flâneuring* through West Edmonton Mall or the Merryhill Centre (for an account of the corner shop's social whirl, see Wilson 1988). From the point of view of the changing urban landscape, in fact, the supermarket is an incredibly important feature, since the policies of store location have had profound impacts on the cityscape, with, for example, the trend towards large peri-urban sites leading to the rerouting of roads and dramatic changes to traffic flows, associated spurts of growth in clusters on the urban fringe, the much-talked-about 'death of the city centre' (and the demise of corner shops, local grocery stores, etc.) and so on. Neil Wrigley's (1996) survey of the changing geography of UK supermarkets gives testimony to their significance as an urban landscape form; it is not accidental that Paul Knox (1991) said that the contemporary cityscape has become the supermarket writ large, such has been the impact of consumption on urban morphology.

By and large, though, with the exception of work by historical geographers, whose emphasis is on the development of urban retailing (e.g. Scola 1975; Shaw 1985), and that of economic geographers, who unpack the links between retailing, consumption and capital (to use the title of Wrigley and Lowe's 1996 volume on 'the new retail geography'), the everyday shopping that occupies most people's time has not been given serious attention, at least in geography (but see ethnographic material by Jackson and Holbrook (1996)).

A notable exception to this trend is Peter Lunt and Sonia Livingstone's *Mass Consumption and Personal Identity* (1992). Subtitled *'Everyday economic experience'*, their study contains a valuable (if brief) section on the supermarket, that most 'common and regular part of our shopping lives' (Fine 1995: 146). Responding to the ways in which supermarket shopping is seen as mundane compared to the consumption spectacles of megamalls, marketing strategies are being developed which 'celebrate the everyday' (Lunt and Livingstone 1992: 97). The provision of exotic produce is central to these strategies, recoding supermarket shopping as a 'high' cultural activity (of course, this applies more notably in 'top-end' supermarkets than in the 'pile 'em high, sell 'em cheap' stores, which are also a significant current retailing phenomenon). And it is also important for supermarkets not to lose sight of their aim: to sell people everything, which must include everyday items; tins of beans alongside sundried tomatoes, toilet rolls as well as deli-counter specials: '[t]he provision of luxuries has to work both in the interstices of the mundane, like day-dreaming, and it has to transform the notion of the everyday' (Lunt and Livingstone 1992: 97). If this mixing is successful, the supermarket should appeal to a broad range of shoppers, all of whom might be looking for different things. Gabriel and Lang's (1995) taxonomy of contemporary consumer identities – chooser, communicator, explorer, identity-seeker, hedonist, artist, victim, rebel, activist and citizen – must all be fed from the shelves if the supermarket is going to reach a sizable market share. And a stroll round any reasonably large supermarket shows how fully this can be accomplished: there are cook-chill ready meals and there are endless raw ingredients, own-label bargains and imported novelties, staples and luxuries.

Every now and then, exposés of the science of supermarkets reveal how controlled the shopping environment really is, in spite of its illusions of consumer democracy. Although these can sometimes read like paranoid fantasies, they give us a valuable glimpse into the organisation of shopping activities. Consider the report by David Runciman (1996) on how supermarkets manage the ways we shop, effectively coercing us into seemingly impulsive purchases (known in the trade as '*splurchases*'): 'We go to supermarkets because they allow us to buy whatever it is we want. And supermarkets make their money by choosing in advance exactly what it is that we are going to want to buy' (Runciman 1996: 2). Technologies called into the service of gathering information about shoppers' wants are usually dressed by stores as being of some benefit to customers – from the 'loyalty cards' which identify and record all our purchases while giving us discounts and freebies, to inventions like 'self-scanning' which, despite its Foucauldian overtones, is sold to shoppers as a handy way to bypass check-out queuing (in the same way as loyalty cards, it also logs all purchases, enabling stores to 'data-mine' and build up a personal profile of each shopper's habits, to use in future direct mailing campaigns and so on). Add to this endless lists of promotions and incentives, plus design tricks (including such well-known psycho-environmental features as the muzak and the smells), and customer aids and comforts (crèches, cafes, bag-packing and car-loading), and it is easy to see why supermarkets make some commentators so worried; the brainwashed shopping zombies of John Carpenter's *They Live* seem closer to the truth than to horror movies.

However, not everyone is so scared of supermarkets, and one of the reasons for this is the realisation that they have a central part in contemporary cultural life. Lots goes on there beyond the buying of biscuits and bananas; as Jan Moir (1995: 5) writes, 'the supermarket is like a soap opera with real soap'. Runciman (1996: 3) goes even further, suggesting that supermarkets are 'one of the few places where entire communities can still meet'. They certainly perform a number of social functions (especially if they have those extra features like cafes and crèches); in Box 5.3 Kate describes the fun lesbians have 'dyke-spotting' in particular supermarkets. And shopping can be a communal activity, even if the conversation is all about the price of fish. Like the cafes of bohemian Paris, today's supermarkets can also be sites of affairs of the heart: one UK chain, acknowledging this, has begun 'Singles Nights' with romantic music, roses and heart-shaped pizza. The intensity and diversity of this social interaction means that futurological predictions such as Internet shopping seem implausible; shopping is not just about shopping, after all.

Box 5.3 KATE

KATE IS A LESBIAN AND ENJOYS THE SOCIAL ASPECTS OF FOOD SHOPPING.

'It is funny going to Sainsbury's . . . I never go without seeing a lot of clients, and also I never go without seeing a lot of dykes as well, which is nice! And people you haven't seen for ages, so you kind of end up – what was going to be like twenty minutes nipping in getting a few things, you know, you end up chatting to people and that's quite – I do enjoy that. It is nice. I love living here because, I mean, partly because there's "Pops" and there's "Zeds" [small local shops; Zed's is a wholefood shop]. And I think there are loads of dykes around here as well and you kind of go round to the local shops and it's nice 'cos again you see people and you might stop for coffee round there. I've never yet been in Zeds when I haven't seen somebody I know. Probably 'cos it's veggie, and it's organic vegetables and stuff.

'I've done a lot of, I mean I don't – I can remember an odd few times when I've gone to Safeway specifically to "dyke-spot" [laughs], oh not specifically, I mean, I have gone to shop, but I mean not just wheel around with an empty trolley. Like I've gone, I've just thought as a change really, I suppose, you know in terms of – and it's usually been on a Sunday for some reason. I don't know why. Em, I wonder why that is? [laughs] If I haven't done the shopping on a Saturday but I know I really need to do some on a Sunday, I do sometimes go to Safeway. So yes, I've heard Safeway is more dykey than Sainsbury's, but I don't really actually know if it is.'

Not everyone shops at supermarkets, of course. There are still those who cannot or will not shop there, whether the reasons are economic or aesthetic (Taylor, Evans and Fraser (1996) provide an interesting account of the variety of shopping practices among populations in two large English cities, Manchester and Sheffield). And, for all their attempts to corner the market in food provision, there are still some culinary ingredients which are not on the shelves, making speciality shops, delicatessens and markets equally important ports of call for the intrepid contemporary urban food shopper. Outside the overdeveloped West, markets often have a central place in the consumer cultures of cities, performing many of the functions which the modern supermarket services (and some which it does not):

> [M]arketplaces are radiating centers. . . . [T]hey are places of gathering, in which virtually all aspects of life are in play, united in shared space. Beyond satisfying their first function as a place of the exchange of goods, they are centers of social life, of communication, of political and judicial activity, of cultural and religious events. They are places for the exchange of news, information and gossip ... In the rich melee of social intercourse the market provides, love affairs are begun, marriages are arranged. Disputes and debts are created and settled, and the market day often ends with a siege of drinking, dancing and fighting.
>
> Buie 1996: 227

This description of the 'dance of exchange', as Buie calls it, shows something of the vitality of consumption in the marketplace, the site where '[m]any bodies, bales, baskets, beasts of all descriptions, all interconnected, [are] all engaged in transactions, big and small' (Buie 1996: 231). For locals, the market is thus a centre of social life. But markets also act as magnets for tourists and visitors, and their appeal is often seized upon in marketing campaigns aimed at seducing travellers in to join the dance. Of course, markets are not only a developing-world phenomenon; supermarkets have not yet managed to erase them from Western cityscapes. The special character of the market – low prices, specialist stalls, the traders' patter – give them continuing appeal, so much so that supermarkets often attempt to create a simulacral market setting for their fresh-produce sections.

EATING PLACES

The business of place promotion is now one of the main activities that cities are engaged in, and their culinary-cultural resources are often appealed to as part of their attractiveness, whether that takes the form of the 'jostling and puddles and honking and smells' (Buie 1996: 231) of the market or the smorgasbord of city eateries. Some cities bear strong associations with a single foodstuff or product – Dundee with its cake, or, in the case of recent advertising campaigns, Prague with Staropramen and Manchester with Boddingtons beers (see Plates 5.1 and 5.2). Other cities trade on the diversity of food and eating experiences on offer – *Time Out*'s 'world on a plate' promotion of London being an example already remarked upon. Urban pride, vested in the gustatory reputation, is central to the cultural capital of many cities: 'Restaurants have become incubators of innovation in urban culture. They feed the symbolic economy – socially, materially, and spiritually' (Zukin 1995: 182). The confluence of urban boosterism and the commercial food business provides many ripe opportunities for the entrepreneurial city around which to self-aggrandise.

Marketing ethnic diversity through food is something cities are particularly keen to do, even though, in a sense, that may diminish their uniqueness. Hannerz (1996: 157) remarks on the transformation of Stockholm's *Hötorgshallen* (indoor market) 'from a fairly traditional Swedish institution into a place where one might shop for Turkish fast food [or] Indian spices'. As cities scramble to become 'world cities', perhaps this globalisation of urban cuisines is inevitable (more on this in Chapter 8). Other cities – or districts of cities – concentrate on more segmented markets; hence, part of Soho's strategy is to appeal to 'pink spenders' (and 'pink eaters') with its proliferation of chic gay bars and cafes (Binnie 1995; Mort 1996). Whether with this kind of targeted niche marketing, or with a globalised 'exotic', or with reinvented traditions of the local, cities are finding endless ways to set themselves apart from each other as players in the game of global cultural-capital chasing.

In the long history of urbanisation, the place which cities have often tried hardest to locate themselves against is the not-urban, the country, the rural. And in terms of food, the country and the city have ambivalent relations. To close this chapter, we want to look at this issue, albeit briefly. Our first stopping place is the Ohio river bottoms of the 1940s, and a paper by sociologist John Bennett (1943) on the ways in which the 'folk-urban' (i.e. becoming-urban) communities of the 'Bottoms' negotiated identities through food. What is interesting about Bennett's detailed analysis is that the urban–rural divide was not simply articulated through culinary practice: city food sometimes had high prestige (examples given include candy, hamburgers, chilli, green peas and fresh fruit), and sometimes low prestige (one interviewee dismissively saying that 'city people are too stingy' with their food (Bennett 1943: 563)). Some food items were coded urban-positive by some people but rural-positive by others (fresh milk being an interesting example). The social stratifications revealed by asking Bottoms dwellers about food preferences suggest an uneven impact of coming urbanisation within communities – much more so than in the fiercely anti-urban Pine Barrens community of contemporary New Jersey, where regional pride articulated through culinary self-sufficiency marks the local store,

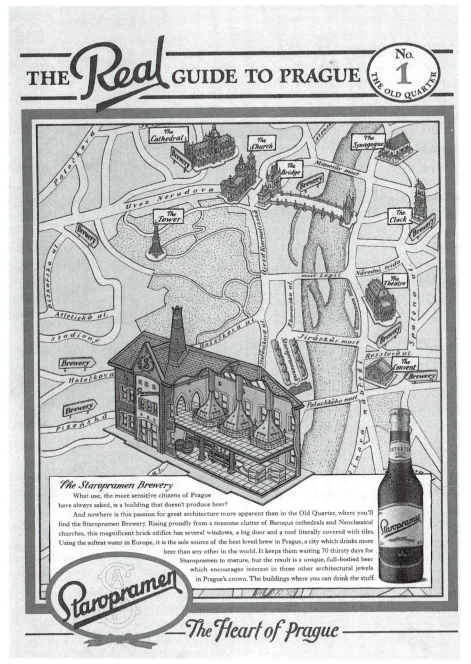

Plate 5.1 Prague: home of Staropramen beer
Source: *Guardian Weekend*. © BBJ/Staropramen Beer

Plate 5.2 Manchester is synonymous with Boddingtons
Photographs: Hannah Taylor. By kind permission of Andrew Williamson

let alone the supermarket, as a forbidden place viewed with hostility by the 'Pineys' (Gillespie 1984). Approaching the same issues from a different angle, Norge Jerome (1980) studied African-American rural-to-urban migrants in 1960s Milwaukee, and found a texture of changes from replacement ingredients to whole new meal forms and formats (including the relabelling of the daily meal cycle from breakfast–dinner–supper to breakfast–lunch–dinner and a decline in seasonal patterning).

The processes of urbanisation, which have impacted profoundly on culinary cultures, have created among many city populations a nostalgia for the countryside, and for the 'plain fare' associated with simple rural life. Fresh foods, 'natural' products, 'home-made' goods are all given high status as rural produce untainted by the negatives of the urban (pollution, industry, artificiality, uniformity) (Lupton 1996). This nostalgia, Massimo Montanari (1994: 159) remarks, reflects 'urban values: the happy countryside is an urban image and one which manifests itself in rural environments only as a reflection (and only if country-dwellers are themselves immersed in urban culture)'. Fads for 'rural' or 'peasant' cuisines revel in the simple earnestness of the country (often essentialised in creations of regional produce, as discussed in the next chapter), yet they hide the economic necessities which created those recipes in the first place. As Montanari puts it, 'What could be more urban than the present-day revival of lesser grains and dark breads? Only a wealthy society can afford to appreciate poverty' (159). And, as is often remarked, the rural–urban divide becomes one of production in the country and consumption in the city: urban dwellers want their farm-fresh veg, but it is country dwellers who must endure the harsh working conditions often associated with putting the goods on the supermarket shelf (Penn and Wallace 1995); 'country food' is provided for city folk, while those who produce it make do with urban-industrial staples.

The city is, of course, a site of many other types of food consumption. It is filled with the activities described in Chapter 3, for example, and it may also be the focus of an articulation of a regional food identity (see Chapter 7). In our discussion here, we have concentrated on some practices which have especially strong associations with the city, and which serve to illustrate our central point: urban identities are routinely enunciated through food habits, and the cityscape is shaped and reshaped by the ever-changing business of servicing those habits. As cities become centres of consumption, food plays an increasingly important role in their cultures.

6

Region

PETER GOULD (23)
Poulet Caribe à la Gould

•

During a period of arduous geographical fieldwork in Jamaica, *Poulet Caribe à la Gould* was conceived in a moment of inspiration while floating quietly on a bamboo raft down a green river to a sunlit azure sea. It must have been Shango. In any case, I poured the rest of my rum into the sand at journey's end as a libation of thanks.

And speaking of libations, two possibilities open up to accompany *Poulet Caribe à la Gould*. The first is a Meursault, say a 1971 Genevrières, Hospise, Cuvée Phillipe le Bon (provided it has been properly cellared), although a 1973 Casses-Têtes, Château Puligny-Montrachet will do at a pinch. Alternatively, Charlie's Rum Shop at Quick Step (District of Look Behind) distils an exquisite white rum (connoisseurs generally prefer Tuesday's), which should be cut 4:1 with lime juice and water to prevent cancer of the oesophagus and perforation of the duodenum. The choice is left to the reader.

But on to . . .

Poulet Caribe à la Gould

First catch your *poulet*, which should be a free-run bird if possible, rather than a plastic-wrapped, battery-produced, hormone-fed, water-injected 'thing'. Split the breast into two portions and remove *all* fat. Turn the pieces inside upside, and with a pointed and very sharp knife remove as much of the white thread tendons as possible. Still inside upside, pat dry, and then, pressing each piece firmly with the palm of the hand, cut horizontally (and carefully!) almost all the way through. Open each piece to a thin butterfly, press and pat dry again.

Take five or six thin green onions and cut in half, discarding the tougher tops. Chop the remainder across. Take a quarter-inch slice of fresh ginger, peel, and chop into fine, almost minced pieces. In a stick-proof frying pan, big enough to hold the butterflied breasts eventually, melt at low to medium heat two tablespoons of fresh, sweet cream butter. Saute onions and ginger *very* slowly (ginger browns quickly) until the white portions of the onions are a light yellow. Remove onions and ginger, reserving as much of the flavoured butter in the pan as possible. Turn to high heat, and when the butter is very hot, but not black and smoking, sauté both butterflied breasts to seal, turning when the first side is cooked halfway through. Upon turning, pour three or four 'squirts' of soy sauce on each of the cooked sides, followed by three or four drops of Tabasco, and a squeeze of a wedge of lemon. Leave the wedge in the pan. Do not overdo the lemon juice, or the dish will be too Sauer. Some find this indigestible. When the pieces are cooked, but not overcooked, reduce heat to low, add the onion–ginger mixture, together with a good slosh of *vin ordinaire*. Continue cooking until sauce is reduced to desired thickness, and serve immediately with rice, fried plantain, salad . . . or what you will.

REGION

Thinking about food and regions throws up some interesting ideas. How, for instance, do certain regions evolve distinctive cuisines (distinct even from their nearest neighbours)? What characteristics of a geographical region impact on its culinary identity (and vice versa)? As regions face an increasingly globalising world, will we witness the erosion of regional distinctiveness, or its reaffirmation?

The definitional question of what makes a region has long vexed spatial scientists, and looking at regions through food can add fruitfully to these debates. As regions seek to market themselves while simultaneously protecting themselves from the homogenising forces of globalisation, regional identity becomes enshrined in bottles of wine or hunks of cheese. Questions of regional uniqueness are thus distilled to the iconic products of particular places. As mobility and hybridity become watchwords of the way the world now works, tradition-bound, defensive articulations of the region may start to look untenable, but the continuing appeal of regionalism, which speaks to us all from the bottom of a bottle of Bordeaux, suggests otherwise, even if – as is often the case – regional traditions are exposed are mere inventions.

6

REGION

•

Sometime in 1995, cola giant Pepsi began an advertising campaign in the UK. A billboard appeared across the country depicting an assortment of faces caught in the act of sipping a cup of cola. This was the latest in a long lineage of 'Pepsi Challenge' ads where consumers were asked to blind-taste two colas and choose their favourite – which, of course, always turned out to be Pepsi. The billboard bore the legend, 'It's make your mind up time.' A few weeks later, the billboards received an addition – across each was pasted an acclamation, a seal of approval: 'Staffordshire chose Pepsi' (Plate 6.1). Here we have a clear ad-agency manipulation of what Yi-Fu Tuan (1974: 101) calls 'local

Plate 6.1 Pepsi appeals to local patriotism in Staffordshire
Photograph: David Bell

patriotism' – exactly the kind of 'patriotism' which is bound to place, and, more precisely, often bound to the region. The billboard offers us a sense of regional pride: we made the right choice, we chose Pepsi.

The region is a spatial scale much beloved of geographers; there is a strong tradition within the discipline of regional studies, and many of the Big Names of geography are associated with work on the region (most notably perhaps Paul Vidal de la Blanche and Carl Sauer). However, with the growth of geography as spatial science, the kinds of study associated with viewing regions as either 'human-centred ecological systems' or as landscapes 'the physiognomy of which was brought about by genetic–morphological processes' was 'dismissed by some geographers with fervour, blame and contempt, while other geographers indifferently let it slip into oblivion' (Hoekveld 1990: 14, 10). Hoekveld's tracing of the development and demise of regional geography contains much of the detail for understanding how regions, as 'mental constructs made by geographers' (13), have suffered changing fortunes in the intellectual ferment of the discipline. His is one of a series of papers surveying regional geography, and projecting renewed interest in the region (Johnston, Hauer and Hoekveld 1990). Such calls for a 'new regional geography' form an interesting platform for considering how the scale of the region is variously articulated through routine practices, including those of food consumption, for the 'regional pride' or 'local patriotism' appealed to by Pepsi (and countless other food producers) often draws on the rhetorics of the region (whether intranational regions such as the culinary regions of France, or international regions such as the Mediterranean with its currently celebrated health-giving cuisine).

In this chapter we will examine selected examples of these regional rhetorics, and explore the ways in which the idea of the region is enunciated through them. One of the important 'research frontiers' for a renewed regional geography highlighted by Johnston, Hauer and Hoekveld (1990) is the cultural identity of regions; as Derek Gregory wrote (1978: 171): 'We need to know about the constitution of *regional* social formations, of *regional* articulations and *regional* transformations.' What better way to do this than by thinking the region through food? (This is hinted at by Nigel Thrift (1990b: 276), who advocates reference to 'Lefebvre's notion of everyday life or Bourdieu's notion of *habitus*' in his calls for a 'critical, theoretical, contextual and polysemic regional geography'.)

In a way, the Pepsi ad is peculiar in its calling to regionalism, since it trades more on an articulation of sameness than the uniqueness usually associated with celebrations of the region: anyone who ventured out of their own region was soon made aware that London chose Pepsi too, as did Yorkshire, Liverpool, Surrey . . . The advert might just as well have said 'Britain chose Pepsi' or even 'The world chose Pepsi'. No billboards read 'Staffordshire loathed Pepsi', of course, or 'London was indifferent to Pepsi'. Stressing uniformity – even uniformity of good taste – seems at odds with the idea of the region, which is more commonly based upon distinctiveness, as we shall see in this chapter.

IMAGINED REGIONS

What we must establish, at the outset, are a few commonsense statements about how the region works in the popular imagination, for it is this use of the region which is deployed in the routine practices of food consumption. So, when Pierre Bourdieu (1991: 222) writes that '[n]obody would want to claim today that there exist criteria capable of founding "natural" classifications on "natural" regions, separated by "natural" frontiers', we would have to disagree at one level – precisely the level of the popular imagination. For example, in Box 6.1 Jack describes his own classifications of North-East foods, and in Box 6.2 Anna, who grew up in London but as an adult moved to Sheffield, recalls her first experiences of Yorkshire foods. And while Phil Crang (1995: 13) is right to remind us that 'regional cuisines are invented traditions', we must also remember the

Box 6.1 JACK

JACK GREW UP IN THE NORTH-EAST OF ENGLAND AND NOW LIVES IN
SHEFFIELD WITH HIS WIFE, LINDA. HE IS RETIRED.

'I suppose our Sunday dinner is, is a Sunday dinner which we've always had, and I, and that involves Yorkshire puddings which are not "à la Yorkshire", you know, 'cos – well, as I understand it, Yorkshire puddings in Yorkshire tend to be large, fat things, whereas Yorkshire puddings in County Durham are small, airy things, and that's the way it's always been done. And I don't like all this business of filled with onion gravy and – that would not be my taste.

'We still, Linda still keeps the tradition of having fish on Good Friday – a strange custom and it's not because there's any Roman Catholicism around, but I mean we always had fish on Friday in the North-East and we still do.

'I suppose one of the main foods of the North-East would be fish and chips because – well it was that part of the North-East which was on the coast and the fish was almost still alive when you got it and it's very difficult to find fish the same now. But I suppose that's where the liking for fish comes from.

'. . . are there local foods in England? I wouldn't go out of my way to have an oggy, for example – a Cornish pasty.

'Again, you see, my idea of a Bakewell tart is the one that you buy from the shops with icing on, whereas Bakewell tarts from Bakewell look more like Yorkshire puddings, I don't like them. And it's an in-built prejudice, you've had what you thought was a Bakewell tart which was acceptable – you have another Bakewell tart and it's not acceptable 'cos it's different.'

Box 6.2 ANNA

ANNA MOVED TO SHEFFIELD FROM LONDON. HER PARTNER, KEITH, HAS
ALWAYS LIVED IN SHEFFIELD.

'The first thing I actually noticed which was really different was chips and curry sauce. In
London I had never seen anyone eating chips and curry sauce and then, I think it was in
my first week of being a student, on the way home from the pub I indulged in them, I
thought they were wonderful. Um, I think that was the strong memory of things being
different [laughing], but since then there is still some things strike me as slightly odd about
– what you see in the baker's is different to what I see at home, the names are different. You
have endless arguments about "are pikelets really crumpets?" and things like that.... On
the baker's stall you'll see things with traditional names and like you'll have your flat bread
cakes rather than your normal rolls, and things like that ...

'Um, another thing I noticed was that they eat their meals a lot earlier, because tea here [in
Sheffield] tends to be around six o'clock or so, whereas supper at home [in London], the
main evening meal would be around eight o'clock.... So I had a bit of getting used to that
there. But now I find generally that the differences are getting less and less because as
Safeway and Sainsbury's seem to take over, there seems to be much less regional food
around, all in all.

'I didn't really encounter regional cooking until I started living with Keith. He's Sheffield
born and bred, um, and most of the cooking which he knows is pretty basic traditional
stuff, so when he came out with his version of Yorkshire pudding, which was totally
different from anything I'd ever seen before, I realised there were perhaps a few differences.
And really heavy and thick and things like onions and herbs in them as well. His ideas of
food are definitely Yorkshire.'

force with which those traditions are invested and reinvested with meaning, and often
with vehement local patriotism. The brilliant case-studies of the Pineys of New Jersey,
mentioned in the previous chapter (Gillespie 1984), and of the Cajuns of Louisiana
(Gutierrez 1984, 1992) show just how important cultural practices like food preparation
and consumption are in marking regional identity; moreover, contra Bourdieu, such
practices centre on an essentialised notion of the region as a 'natural' space – as favoured
in humanistic geography's work on the centrality of 'place' to identity formation (Thrift
1991). The region is often seen by those evoking a regionalist discourse exactly as
'natural' in every sense of the word. The region is also likewise seen as in some senses
a 'local' space – often to be set against the (homogenised) global (a point we shall return
to later). While it is important for us to deconstruct such (often romanticised, equally
often politicised) specifistic notions of the region, we must not lose sight of their
significance in everyday articulations of regionalism, such as those based around food

and eating. Johnston (1990: 130) makes a key point here:

> The creation of regions is a social act. Regions differ because people have made them so. Undoubtedly in many cases differences in the physical environment will influence the creation of regional variations, with different environmental conditions stimulating different individual and social responses. . . . But similar physical environments can be associated with very different human responses, and similar patterns of spatial organisation can be found in very different milieux.

The region, then, is a product of both human and physical processes: a natural landscape and a peopled landscape. It is also a powerful way of thinking place and identity (Thrift 1991). It may be, as Bourdieu (1991: 223) asserts, a *'performative discourse'* in which the region is called into being by being named and claimed, but, as we know from other arenas in which essentialism is critiqued and identities revealed as 'fictions', the ways in which people make sense of their lives are often at odds with academic theorising (Fuss 1989).

The example of the 'wine region' is an exemplar of how the rhetorics of the region are pulled together to stress uniqueness, and how the physical and human landscapes are seen as together producing that uniqueness. In the Introduction we briefly discussed the concept of *appellations d'origine* and how they are used to embody regional uniqueness, acting as a kind of 'trademark' and quality guarantee for wines (Moran 1993). The argument used is straightforward: wine is made in a particular place where environmental conditions mixed with the (inherited, traditional) artisanal production techniques produce a distinct character for a wine; for consumers, this means that they know all they need to know about the provenance of their bottle of wine, the conditions and processes of its production, and therefore the expected qualities it should have (if they have the knowledge, that is). Wine tasting is based on this very idea, as is wine advertising: three newspaper adverts for Piat Père et Fils wines stress the character of the soil, the role of tradition and the attitude of the producers in making their wines distinctive (Plate 6.2). Producers have long been hostile to the 'abuse' of this attachment to place inherent in wines (German 'champagne' or 'Spanish Chablis' being therefore geographical impossibilities; more recently, a British manufacturer was prohibited from naming a sparkling cordial 'elderflower champagne') (Unwin 1991). Legal demarcation of vineyards and wine regions through *appellations* and other similar markers thus creates a series of monopolies on who can produce which wines, and where – more fiercely guarded than ever in today's global wine market.

Soil, as any self-respecting Frenchman or woman inherently knows, requires the very same
tender loving care that's so selfishly lavished on the vines themselves.

Such unflinching devotion, they also know, will be repaid a hundred fold by the superior quality of the wine produced.

Hence Piat Père et Fils' zealous application of their expertise in creating fine wines that dates all the way back to 1849.

For that self-same, self-respecting Frenchman or woman wouldn't even deem to partake of Boeuf Bourguignon, or Steak au Poivre,

were it not enhanced by the full, blackcurrant bouquet of Le Piat de Cabernet Sauvignon.

Produced from one of the noblest varieties of grape, it's a wine some people would give the earth for.

Others, thankfully, even give up their beds.

WHAT ON EARTH MAKES

A PLEASURE LOVING FRENCHMAN

GET UP AT 3AM FOR WORK?

THE EARTH.

Plate 6.2 Local soils and regional traditions create the distinctive taste of Piat Père et Fils
Source: *Guardian Weekend*, July 1995. © Burkitt Edwards Martin Ltd/Piat Père et Fils

PRODUCING REGIONS

The idea of *appellation*-like geographical product indicators is one which has spread far beyond vineyards to become a common marketing strategy in contemporary food consumption (Hodgson and Bruhn 1993). Almost any product which has some tie to place – no matter how 'invented' this may be – can be sold as embodying that place. Bottled mineral water is one product often intimately tied to place – to the exact location of the source, which brings with it notions of purity and naturalness. Labels on bottles of mineral water often include diagrams or maps of springs and aquifers, together with sketched histories of water-taking at source (the Victorian fervour for spa waters having, in some cases, established a 'tradition' based on the same characteristics of the water used to sell it today). Despite the rise of down-market generic bottled waters not tied to source, the use of locality to ensure product quality maintains the upmarket bottled waters' dominance on supermarket shelves.

The case of cheese is similar to that of wine and water, with products such as 'Somerset Brie' contesting the strength of the spatial association of this particular soft cheese with the Île-de-France region. As Moran (1993) notes, French cheese producers have responded to such innovations in two different ways: either by tightening the definitions of cheeses (specifying the breed of cows whose milk must be used, and the feeds appropriate for those cows) to exclude other producers, or licensing and endorsing other producers (providing, for a fee, the expertise to emulate production in the home region). While the latter strategy emphasises the human input to cheese production over the role played by the physical landscape (thus differentiating it from the stance of wine-makers or water-bottlers), it maintains the quality stamp demanded by customers.

Part of that quality stamp is also the guarantee of things which have *not* gone into the production process – a certification against adulteration, with the consumer able to enjoy the product safe in the knowledge that, thanks to the combination of environmental conditions and carefully regulated 'craft' production processes, they know precisely what they are eating or drinking. This guarantee has itself been generalised over whole areas of food production, most notably fruit and vegetable growing (but also animal husbandry), into the notion of organic foods. As an antidote to what many see as the sterility of the supermarket, there has been a blossoming of organic suppliers, overseen in the UK by the Soil Association – maintaining a fundamental link to the physical landscape (Blythman 1995; James 1993). Coming along as part of a growing consciousness about both environmental and health concerns over food production, organic food reinstates some of the attributes associated with wine regions: the labelling of food as organic tells consumers all about the conditions of its production (small-scale, chemical-free, non-intensive, locally sensitive, countercultural, etc.).

Such a commingling of associations finds further refinement in macrobiotics, which stresses a complete 'philosophy of diet' embracing, among other things, eating foods grown in the climatic conditions in which consumers live, respecting the seasons, and eating from one's immediate locale:

> A macrobiotic diet is . . . based as much as possible on regional foods, especially with regard to fruits and vegetables. This preserves the harmony between plants and environment that automatically develops in any particular region, a harmony that should be reflected in our diet.
>
> Heidenry 1992: 66

While macrobiotics remains a fringe dietary habit, there have been significant transformations in the ways we eat based upon notions of particular regions. Perhaps the most notable of these has been the widespread promotion of the so-called Mediterranean diet (Cannon *et al.* 1994). The characteristics of the Mediterranean diet – simple 'peasant' food, with emphasis on pastas and breads, grilled fish, fresh fruit and vegetables, red wine, and especially olive oil – have been celebrated in the UK as far healthier than our own diet (lower in cholesterol, lower in saturated fats and dairy produce, lower in sugar, high in antioxidants), leading to advocacy of a Mediterranean-like diet in the government White Paper *The Health of the Nation* (Department of Health 1992), plus extensive promotions of Mediterranean fare by supermarkets, as well as innovations like olive oil spreads (a margarine-like butter substitute). The longevity of the Mediterranean peasantry is re-created in advertising for such products (see Plate 6.3), while the lineage of our love affair with Mediterranean food traces back to the legendary food writer Elizabeth David and to the boom in tourism to the region (Hardyment 1995). Of course, the idea of a distinctive 'Mediterranean diet' hides national and local variation, which is in some cases quite marked, offering instead a condensation based around a few items drawn principally from the southern Italian table (those listed above), as well as glossing over changes in the diets of Mediterranean countries: Crang's 'invented tradition' seems a most appropriate tag to apply here, especially when one considers the history of the Mediterranean diet, which stretches out across the globe (Delamont 1995).

INVENTED REGIONS

Such a glossing of regional differences into the diet of a single international region – the Mediterranean – runs contrary to the strong tradition of classifying a number of distinct culinary regions within any one country. The cuisine of France, for example, is associated very strongly with a series of reasonably distinct regions where different physical

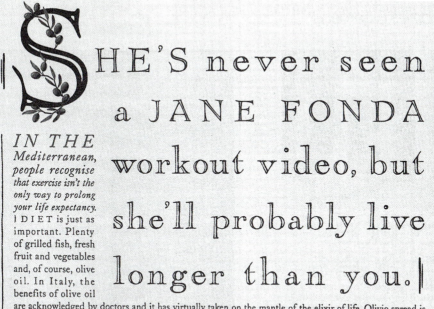

SHE'S never seen a JANE FONDA workout video, but she'll probably live longer than you.

IN THE Mediterranean, people recognise that exercise isn't the only way to prolong your life expectancy. DIET is just as important. Plenty of grilled fish, fresh fruit and vegetables and, of course, olive oil. In Italy, the benefits of olive oil are acknowledged by doctors and it has virtually taken on the mantle of the elixir of life. Olivio spread is a delicious way to incorporate the oil from age-old olive groves into your daily diet. Simply use it to take the place of saturated fats like butter (which can actually increase your blood cholesterol levels) and immediately you'll be reducing your saturated fat intake. Of course a Mediterranean diet alone won't give you the health and figure of a supermodel. But it may add a few years to your life. For more information on Olivio spread and the Mediterranean diet, phone free on 0800 616030.

OLIVIO SPREAD. PART OF YOUR MEDITERRANEAN DIET.

Plate 6.3 The advantages of a Mediterranean diet
Reproduced by kind permission of Olivio

conditions combine with localised culinary histories to produce a rich map of food custom, habit and practice (see Figure 6.1). These gastronomic regions include the Basque Country in the Pyrenees, which is renowned for its mountain-cured hams and *confit* (goose, pork and duck, salted, cooked and preserved in their own fat); Brittany on the Atlantic coast, which is celebrated for its mussels, lobsters and oysters, and dishes such as *palourdes farcies* – clams served in shells with a *gratine* stuffing; Alsace, a fruit-growing region which is famous for its jams, preserves and fruit liqueurs (especially the Framboise); and Lorraine, a dairy-farming region in north-eastern France which has given us the quiche (David 1960; Floyd 1987). Italy is similarly marked by regional patterning, and the homogenised Mediterranean diet described above would be unfamiliar to many inhabitants of the north of the country, where meat, dairy produce

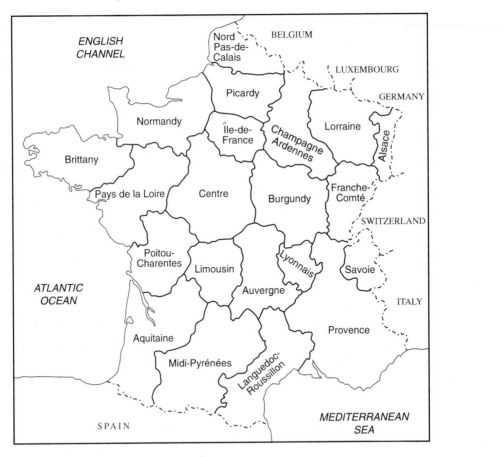

Figure 6.1 The gastronomic regions of France
Source: Redrawn by Graham Allsop after Eugene Fleury

and rice are much more central elements of the diet.

Elisabeth Ayrton's (1980) *English Provincial Cooking* lays out a set of distinctive cuisines for the following regions: East Anglia and the eastern counties, the North, the Midlands, the West Country, the South and South-East, the Home Counties, and London. Many of the dishes listed bear more specific place-markers: Huntington stuffed pears, Lancashire hotpot, Melton Mowbray pie, Devonshire splits, Sussex pond pudding, Greenwich whitebait and Yorkshire pudding. Describing the last of these dishes – a pancake-type batter cooked in boiling fat in the oven – Jennifer Stead nicely captures the way particular foodstuffs are seen to embody regional characteristics:

> Yorkshire pudding is tarred with the brush of the county character. Yorkshiremen were notorious for their sharp practices, and for centuries to 'have Yorkshire put on you' or to suffer the 'Yorkshire bite' was to be cheated. At the same time Yorkshire hospitality was fabled. So Yorkshire pudding is a triumphal marriage of those conflicting aspects of the Yorkshire character – meanness and liberality. It is cheap and hugely filling ... the fact that they [Yorkshire puddings] require spanking hot fat, explosions as the batter hits it, fierce heat, and crisp results, may explain why it has often been said that only Yorkshire folk – those possessing the Yorkshire temperament – can make a true Yorkshire pudding.
>
> Stead 1991: 146, 150

Although Ayrton's and Stead's stress is on history and tradition, such regional specialities and specialisms continue to flourish, now with added place-promotion possibilities and inter-regional rivalries. Thus, the creation of the West Midlands as home of that newest edition to ethnic eating in the UK, the balti from northern Pakistan, has involved setting the region up as the authentic British base of balti cuisine, with Southern (especially London) balti houses dismissed as pale imitations (accused of merely serving up standard curries in the characteristic wok-like balti or karahi). Baltis have been served in Birmingham for twenty years – long enough for one Midlands balti *aficionado* to dismiss London's balti houses as 'trading on Birmingham's heritage' (Tredre 1995: 11).

A brilliant case-study of a similar kind of investment in regional cuisine exemplified through a single form is Timothy Lloyd's paper 'The Cincinnati chili culinary complex' (1981). As he states at the opening of his paper, in terms which could equally apply to the balti debate in England,

> It is common to read that developments in food processing, refrigeration, and transportation, which have made foodstuffs from any region of the United States available throughout the country and throughout the year, have contributed to the demise of regional foodways. The availability of a great number of

foodstuffs, however, does not in and of itself guarantee the elimination of traditional food habits.... [Further], it is possible for new foods to be introduced, or even to be invented, and then over time to have their own system of custom and identification build up around them.

(Lloyd 1981: 28–9)

In the case of Cincinnati, chili (boiled minced beef with a mixture of herbs and spices – similar to 'chilli' in the UK, which is usually served as chilli con carne, with kidney beans added) has been incorporated, modified and reinvented to become the core of a region-specific cuisine with its own intricate rules and patterns, into which is invested considerable regional pride. One interesting feature that Cincinnati chili shares with the balti is that, unlike many regional foods, it is not based on locally derived ingredients or 'traditional' food habits. Like the balti, it began with commercial enterprise (starting in this case with the arrival of one Tom Kiradjieff from New York City, who started making chili to sell from his Empress Coney Island stand). From these humble beginnings evolved the whole 'culinary complex' with its variations, or 'ways' of eating chili (for example, chili on a bowl of spaghetti with grated cheese is known as a 'three-way'; add to that chopped onions and kidney beans and you have a 'five-way'; chili on spaghetti with cheese but no onions is not a 'three-way plus beans' but a 'five-way no onions', and so on). In a second similarity to the balti story, the Cincinnati chili represents an Americanisation of ideas culled from Kiradjieff's background in Macedonia – or, more precisely, a 'Cincinnatification', just as baltis are perhaps a 'Brummification' of Pakistani cuisine; each, by adapting to suit generalised local tastes, established its position as a regional dish similar to – but still unique from – those set at the nation's table.

Both the balti and the Cincinnati chili have a final common feature which sets them apart from the ways some regional foodways operate: their commercial focus means they are made to appeal to a wide audience; part of the regional pride invested in them comes from selling their particular regional taste to outsiders. Some regional foods, however, may be used to set 'locals' apart from non-locals – as culinary boundary markers. C. Paige Gutierrez's work on Cajun cuisine from Louisiana shows this: staples in Cajun cookery such as crawfish, boiled crab and boudin (a kind of sausage) are relished by Cajuns but are often seen as virtually uneatable by outsiders; they are difficult to eat because they require considerable knowledge and skill, and the spices used render them intolerable to those with unaccustomed constitutions (Gutierrez 1984, 1992). When Cajun food is marketed to outsiders (through restaurants for example), it is thus often 'watered down', much to the affrontery of the radicalised Cajun community of 'Proud Coonasses' (although other groups, such as the 'Genteel Acadians', are more positive about the export possibilities of Cajun cuisine, which is a significant economic boost to the region through tourism, leading to what Gutierrez (1992: 128) calls a 'loosening of boundaries' of regional identity marked by food).

What the examples used throughout this chapter share is a conception of the region as 'local space' – set against forces of globalisation (or asserting itself within them, but only under conditions which ensure the maintenance of regional identity). They also share an agenda which could be labelled the commodification of the region, which is part of the same project, often tied, as has been emphasised, to an essentialised notion of what constitutes the region – its selling point. If, as Nigel Thrift (1993: 97) asserts, '[t]he new regional geography is about mobility', then selling the region is one way of 'mobilising' regional distinctiveness in an age of global consumerism. Thrift's comments, which open and close one of his contributions towards rethinking regional geography, seem appropriate here for concluding our trip round the regions, and leading into the subsequent chapters, on the nation and the global:

> The world is becoming a more interconnected set of places as the internationalisation of economies, societies, and cultures goes on apace. . . . This is not to say, however, that regional diversity is dying out. Local difference can still be important, indeed may even have become more important, if in different ways. . . .
>
> The global systems of today can reproduce regions with distinctive economic, social, and cultural structures. . . . Certainly the relationship between the global system and regions is now based on a more complex web of interconnections than ever before.
>
> Thrift 1990a: 180, 204

7

Nation

JOHN AGNEW (47)

●

4 cloves garlic
1 packet fresh (or frozen) leaf spinach
4 medium-sized potatoes
1 tablespoon chopped parsley

Chop garlic and lightly brown in ¼ inch of olive oil in a heavy skillet. Take previously soaked fresh or defrosted frozen spinach and mix with olive oil and garlic. Take previously boiled (and skinned) potatoes, chop them and place in skillet. Add parsley to the mixture, stir, and cook over low heat, turning frequently. Serve, add pepper to taste.

This meal tastes better after pasta (and light sauce) and if eaten in Italy.

NATION

What is archetypally American food? What items constitute the British diet? The longer we ponder questions like these, the more difficult they become to answer. National cuisines are in a process of constant reinvention, absorbing new influences and letting some traditions die out. Nevertheless, a nation's diet can have a key role to play in nationalistic sentiments, with threatened invasions of 'filthy foreign food' being seen as dangerous to the whole fabric of national identity.

At the same time, national cuisines other than our own are often celebrated, even fetishised, for their exotic difference, adding spice to life. The histories of individual foodstuffs, like the histories of migrant groups, thus tell tales of both xenophobia and neophilia, of dread and desire. Sometimes a nation's identity is captured by a single food item. In Britain, curry is synonymous with India, for example. In other cases, it is the manner of cooking as much as the ingredients and final dishes which define a national cuisine (as might be argued is the case for French cookery).

The nation is also an organisational unit, of course, and the bodies which regulate its behaviour always take an interest in what it is eating. Monitoring and modifying the nation's diet can be a central administrative task for government, since so much depends (or could depend) on the food provision and consumption of the population.

7

NATION

•

It was an interesting experience to be in the USA when the latest and biggest wave of bovine spongiform encephalopathy (BSE), or 'mad cow disease', hysteria broke in Britain, and subsequently around the globe, leading to a worldwide ban on British beef (Plate 7.1). The US news media covered the events of March–April 1996 with lurid interest, especially when American fast-food giants like Burger King, Wimpy and McDonald's all stopped selling hamburgers in their UK branches, pending the sourcing of non-British beef. For a few days, to be English in Las Vegas was no longer to be complimented on one's accent or asked about the Beatles; it was to be part of a whole new spurt of food nationalism, with Mad Cows and Englishmen the object of mild ridicule from our American hosts (whose beef was upheld – and devoured – as definitively safe and wholesome). Back in the UK, the BSE drama unfolded in part as another episode in the unending Britain-versus-Europe soap opera (a joke even ran among the meat and livestock industry that BSE actually stood for 'Bloody Stupid Europeans'; see Chapter 4 for more anti-Europeanisms), generating the usual U-turns in policy and illogical commentaries from food, medical and animal specialists (one of the country's veterinary societies apparently missed the contradiction in its declaration that the wholesale slaughter of beef cattle would be 'immoral', when all that those cows are bred for is precisely to be slaughtered).[1] Whatever the scientific 'facts' of BSE, its route into beef cattle and its links with CJDV (Creutzfeldt–Jakob disease variant, known as 'the human form of BSE'), and whatever the real costs of the mad-cow panic (lost farming and food retailing revenue, bankruptcy in associated industries, a projected tourist slump, even suicides), its position at the locus of discourses of food nationalism – especially considering the long-standing emblematic status of British beef on the nation's table – serves as a neat example of the core themes of this chapter.

As Peter Bishop (1991) notes, the association of Britishness with beef has a long history, and has frequently been called upon at times of national crisis, when it becomes counterposed against what Dena Attar (1985) calls 'filthy foreign food'. One particularly reviled national cuisine is French: Bishop quotes an eighteenth-century commentator,

Plate 7.1 'Mad cows and Englishmen'
Source: Reproduced from the *Guardian*, 5 June 1996. © Steve Bell

who wrote scathingly of food 'dressed after the *French* fashion', saying that 'Fish, when it has passed the Hands of a *French* Cook, is no more Fish.... It, and every Thing else, is dressed in Masquerade' (quoted in Bishop 1991: 32). A more recent but no less Francophobic example, from British tabloid newspaper the *Sun*, is quoted by Deborah Lupton (1996: 26), in which the paper ranted about 'dodgy food', 'garlic breath', 'foul soft cheese . . . riddled with listeria bugs', and (somewhat ironically) 'false' French claims about BSE in British beef. The nationalistic meanings of beef account for a large part of what Nick Fiddes (1991) calls 'the power of meat' for Britons.

In fact, food and the nation are so commingled in popular discourses that it is often difficult *not* to think one through the other; as Bishop notes, 'diet's relationship to cultural identity is not just a jingoistic one . . . [but is] on a par with language in terms of cultural definition' (1991: 32). Like a language, food articulates notions of inclusion and exclusion, of national pride and xenophobia, on our tables and in our lunchboxes. The history of any nation's diet is the history of the nation itself, with food fashions, fads and fancies mapping episodes of colonialism and migration, trade and exploration, cultural exchange and boundary-marking. And yet here begins one of the fundamental contradictions of the food–nationalism equation: there is no essential *national* food; the food which we think of as characterising a particular place always tells stories of movement and mixing, as 'deconstructions' of individual food histories, such as those in Sara Delamont's *Appetites and Identities* (1995) or Margaret Visser's *Much Depends on Dinner* (1986), show. If, as Benedict Anderson (1983) has famously proclaimed, the nation is an 'imagined community', then the nation's diet is a feast of imagined commensality. It is ironic, then, that those who claim to hate 'foreign food' and eat only 'plain old English fare' fail to realise that there is no such thing; all there is is a menu of naturalised foods brought to Britain's shores through the course of history, modified, adapted and hybridised over time (British food is described by seven different voices in Box 7.1). We all know, but routinely forget, for example, the origins of the potato, that mainstay of the British plate (just as we ignore the Egyptian or Spanish sourcing of the early new potatoes we relish as a first taste of the English summer). Furthermore, the foodstuffs we think of as definitionally part of a particular nation's sense of identity often hide complex histories of trade links, cultural exchange, and especially colonialism. Stuart Hall (1991: 48–9) summed this up perfectly in the context of the black presence in Britain as read through that great signifier of Englishness, the cup of tea:

> People like me who came to England in the 1950s have been there for centuries; symbolically, we have been there for centuries. I was coming home. I am the sugar at the bottom of the English cup of tea. I am the sweet tooth, the sugar plantations that rotted generations of English children's teeth. There are thousands of others beside me that are, you know, the cup of tea itself. Because they don't grow it in Lancashire, you know. Not a single tea plantation exists within the United Kingdom. This is the symbolisation of English identity – I mean, what does anybody in the world know about an English person except that they can't get through the day without a cup of tea?

Despite this colonial contribution, which makes a nonsense of English food patriotism, lamentations of the loss of Great British Grub continue to come thick and fast, from academic and lay commentators. Bishop cites Raphael Samuel:

> There is no such thing as a national diet [these days], as there was in the days when 'Boiled beef and carrots' provided a rousing music hall chorus, and 'solid breakfasts' and 'suet puddings' could be taken … as the distinctive signs of English stability.
>
> Samuel 1989: xxxi, quoted in Bishop 1991: 36

Box 7.1 BRITISH FOOD, IN SEVEN VOICES

'I would say something like a steak is British, traditional Sunday dinner's British, fish and chips is British, from the chip shop. Um – tends to be virtually everything like that. I would say foreign food's pasta, and after pasta you're looking at pizza, is that foreign or American or is it Italian, 'cos, er, pizza's supposed to be from Italy but is it more from America now?'

'So it's very rare that I go and cook an English meal because I find it right, you know, to cook an English meal at home it's like grilling up fish fingers and warming up peas and some vegetables or something like that, I find it right boring and plain, right boring and plain.'

'British is your sausage and your, your joints of beef or – just your traditional meat and two veg, I would have thought as British, or stews, pies, that type of thing. Maybe chilli, or lasagne, or spaghetti bolognese is foreign, Chinese I'd say was foreign.'

'Yes I think there is a British cuisine and unfortunately I think it's now to be found in an awful lot of pubs which have signs outside saying "good food". You know, they started off this sort of pub eating and a lot of it was very good but everyone has jumped on the bandwagon now and I think if you really want to sample the worst of British food then just go to the average British pub – overboiled vegetables again. You come back to these carrots which are wishy-washy and tasteless, and overboiled spuds that have been heated up. I don't think the old British food was very good.'

'I mean it would be an enormous fried breakfast and then it would be, actually I suppose traditionally it would actually have been probably more meat than there is today. Beef or lamb or chicken, something like that, and probably the heavy puddings.'

'British food? Meat and two veg, in its most boring way served up by your elderly relatives. With potatoes, whole, peeled, boiled to goodness knows how long – yes, and some very limp cabbage. Yeah, that's a tricky one, because I have a problem with that; whenever I'm entertaining, oh, you know, foreign students or something, which isn't often but I have had friends here from Hong Kong and places like that and I've wanted them to come round for a meal and give them something typically British. And it's a bit limiting when you're vegetarian and you want to try and think of something that's vaguely British in style but is vegetarian. Um, so I suppose, I don't know, something like a cottage pie but made with lentils would be one way of getting round it. And I think, sort of, stodgy puddings and cakes are typically British as well.'

'I suppose most people would think of, like, the steamed puddings, wouldn't they? More or less kind of like school dinnery-type stuff, but I think British cooking's opening up a lot more. There's not a great deal that you'd pick out as being British, other than the Sunday lunch with Yorkshire puddings, cooked breakfast – that's what you traditionally think of as British food. I'm sure there's lots more to it than that. Well, I do like doing it, it's just such a faff.'

Writing in a similar vein in the *Observer* newspaper, novelist and food critic Will Self finds the very definition of Englishness in the disappearing 'traditional' full English breakfast:

> People often wonder nowadays, as our sense of nationhood is whittled away, what it really means to be English. I would like to advance a challenging definition. It's not a question of custom, belief in given institutions or a willingness to mull over the minutiae of precipitation. Oh no, it's the full English [breakfast]. If you eat a full English first thing in the morning – you feel English.
>
> Self 1995a: 32

And in a final example, Matthew Fort (1996a: 31) comments that '[t]he decline in the number and quality of our native [food] products should be almost as great a matter of concern as the destruction of the rainforests', arguing that government policy is failing adequately to protect our indigenous cuisine, and that British cooking remains uninspiring compared with that of some of the country's neighbours. However loudly he cautions against romanticising and nostalgising some lost golden age of British food, Fort, like Self and Samuel, tries to tie down national culture with certain traditional culinary forms and practices – to find an 'authentic' British (or, more often, English) cuisine.

MAKING AND REMAKING NATIONAL CUISINES

Of course, a substantial part of this lamentation is aimed against the American cultural imperialism of fast foods, almost universally targeted as the prime suspect in the murder of Great British Grub. But it is not just in the UK that McDonaldisation is singled out as deforesting national cuisine; Rick Fantasia (1995) has written a detailed paper on what he describes at the outset as a culinary oxymoron: fast food in France:

> Fast food, with its suggestion of speed, standardisation, and the homogenisation of taste, would seem to represent the direct inverse of French gastronomic practices. . . . [A]nalysts of French society and of American popular culture have considered France virtually immune to the spread of standardised eating practices generally, and fast food in particular.
>
> Fantasia 1995: 202

Despite this supposed immunity, fast food is not considered merely an unthreatening irrelevance in France; it is seen as dangerous and predatory, and its continued predation led in 1989 to the establishment of a branch of the French Ministry of Culture dedicated to national 'culinary arts'. Fantasia is quick to emphasise, as we do here, that French culinary culture is the product of a 'several hundred years of cultural change, appropriation, expansion, and domination. What appear to be indigenous foodways', he continues, 'may actually be cultural imports, like the *café*, whose French cultural identity is relatively recent, having been imported in the seventeenth century' (Fantasia 1995: 205). That said, the popular currency of ideas of national culture (including culinary culture) makes the investigation of food and nationalism an important exercise.

Fantasia's account of the social space of fast food in France makes fascinating reading, but maps familiar terrain in terms of the processes of acculturation which occur when new foodstuffs enter into a nation's consciousness. Prior to the 1980s, he notes, hamburgers were unknown except to the fashionable elite, served in swanky restaurants in Paris but unknown to the masses. So trendy was the hamburger, in fact, that a fashion designer chose McDonald's to provide catering for his Paris show. The chic appeal of American junk food was, at least in part, as a transgressive rejection of the seriousness and *Frenchness* of French culture. As with many contemporary food fads, the French adoption of the hamburger began among the style-conscious new service class, who use food as a marker of social position, or distinction in Pierre Bourdieu's (1984) terminology. The burger in France, as elsewhere, is, moreover, the favoured food of the young: Fantasia reports that almost four-fifths of paying customers in a survey in 1989 were under 34 years of age; even as fast food percolates down the social strata, it remains a predominantly youthful taste. In interviews, the young burger-eaters confirmed a love of the Americanness of fast-food outlets, which they contrasted with the 'traditional' French cafes they saw as less appealing. Multinationals like McDonald's have capitalised on this, by establishing burger outlets which play up their American difference (despite making some interesting minor cultural adjustments, like abandoning standard-issue fixed seating in favour of movable chairs more akin to those found in cafes, and – to recall Vincent Vega's discussion which opened this book – selling beer with burgers). As Fantasia (1995: 235) concludes:

> the emergence of the fast food experience in France [was found to be] culturally and socially decontextualised. For the industry, the workers, and the consumers, this has been precisely the point – that fast food has embodied what have been regarded as distinctly 'American' practices, offering the taste of an 'Americanised' world.

A second study – in some ways very different, in others very similar – of the ways in which the 'taste-makers' in society, the new middle classes, make and remake a national cuisine is found in Arjun Appadurai's 1988 article on Indian cookbooks. Appadurai's approach is to 'view cookbooks in the contemporary world as revealing artefacts of culture in the making' (1988: 22), and particularly to see how 'Indian' food is represented to an Indian market specifically urban, Anglophone and postcolonial – a 'polyglot culture' which is multiethnic, multicaste, cosmopolitan and Westernised (a group which, like its French counterparts, also enjoys fast food (Narayan 1995)). The representational strategies of cookbooks aimed at this market are quite complex, since they seek to combine some constructed notion of a national cuisine with a sensitivity to 'authentic' regional and local differences, while taking into account the gastronomic traditions not only of Hindu and Islamic cultures, but also of colonial India. A common strategy to manage this culinary diversity is through the construction of staged menus, which combine to give an elaborate 'taste of India'. While these snackshots of Indian cultural diversity rely on metonymic 'ethnic cameos', with a whole region's culinary traditions condensed to a single dish (often one, Appadurai notes, which would probably not be seen as 'typical' or even necessarily recognisable by an 'insider' from that region), these menus are perhaps more remarkable for the nationalist ideologies they carry, which stress precisely the twin themes of rich variety and cultural unity.

Appadurai's article makes for interesting reading in the context of the representations of Indian cuisine in Britain, where the colonial legacy combines with the presence of a population from the Indian subcontinent which has established its own culinary cultures there. Perspectives on the state and status, as well as the significance, of 'Indian' food in Britain vary considerably. For Christina Hardyment (1995: 125), the 'strengthening individual identity of Asian restaurants' in the UK shows that 'the answer to coping in a multicultural society is celebration of difference rather than obsessive integration; informed respect rather than grudging tolerance'. From a much more critical perspective, Uma Narayan (1995: 64) uses curry to think about a whole range of issues in Asian cultural politics, considering the full complexities of Indian food in the colonial and postcolonial histories of both India and Britain. Her nuanced discussion offers interesting critique of those theorists who rail against 'eating the Other', for while she is unimpressed by the *Indianness* of food served in British (or American) Indian restaurants, she retains a critical balance between consumption and production, always weighing the economic importance of the food sector to the Asian community against her distaste for the 'culinary imperialism' lying limp on her plate. At the same time, she avoids over-romanticising the contribution of the ethnic food business, reminding us of the low-cost, low-profit, long-hours conditions common in the sector. Rhetorically answering the question of 'ethical eating' in response to her critique, Narayan offers some tentative suggestions:

> I would argue that western eaters of [non-western] ethnic foods need to cultivate more reflective attention to complexities involved in the production and

consumption of the 'ethnic foods' they eat. They might, for instance, reflect on the race and class structures that affect the lives of the workers who prepare and serve that food, and on the implications of class differences between the immigrants who own these restaurants and the immigrants who work for them. They might think about the fact that while low-cost, 'mainstream' eating places, such as diners and fast-food chains, employ a predominantly female labour-force, many 'ethnic' restaurants employ mostly male immigrants. . . . They might reflect on the fact that some of the same restaurants that provide them with cheap and 'exotic' ethnic food often serve as regular eating-places for 'ethnic' male immigrants away from their families, and as social meeting-places for them.

Narayan 1995: 78

Hardyment's tidy history of ethnic foods in Britain, in *Slice of Life* (1995), has some nice detail on how 'Indian' restaurants (predominantly owned and run by Bangladeshis from a region called Sylhet) were started by the male-only immigrant population to serve their own needs, and only began to court white customers when immigration laws were changed in the 1960s to allow Asian families into Britain (although there had been high-class Indian eateries in Britain since at least the 1920s, serving nostalgic old colonials upon their retirement). Interestingly, Hardyment singles out students as the group responsible for popularising both Indian and Chinese food in the 1960s, lured by its cheapness and exoticness when compared with refectory stodge or their own first attempts at gastronomy. They in turn paved the way for a generation of 'lager louts' to make their weekly pilgrimage at pub-shut time for chicken vindaloos all round – a tradition which continues (among students, too), and which surely represents one of the most ambivalent and troublesome of the production–consumption relations Narayan urges us to think about. Later diversifications, such as tandoori, and most notably in recent years balti restaurants, together with take-aways, ready-cook meals to heat and serve at home, the wider availability of ingredients and the publishing explosion of cookery books, have given 'Indian' food a crucial place at the very centre of Britain's national diet. British TV company Channel 4 even ran a situation comedy based upon an Indian restaurant, the critically unacclaimed *Tandoori Nights* (Daniels and Gerson 1989), while Indian supercook Madhur Jaffrey has done much to popularise Indian cooking through her many books and TV series.

The representations of Indian food in the British food media make for reading as interesting as those written for the domestic market discussed by Appadurai. Two brief examples will suffice here, each from a very different perspective, but both dealing (more or less explicitly) with 'colonial India'. The first is Harvey Day's *The Complete Book of Curries*, published in 1966. A quotation from the Foreword sets the tone of the whole volume:

The influx of Indians, Pakistanis, West Indians and other curry-eaters from the Commonwealth into Britain within recent years has seen the sprouting of Indian and Pakistani restaurants like mushrooms in a midden. Which should cause the non-curry eating public to pause, for the popularity of any cult or fashion always brings in its wake a rapacious horde eager to leap on the band waggon.

All who run restaurants where curries are offered to a gullible public are not experts in their native art and the result some achieve on their patrons is a revulsion to curries of every sort. These restaurateurs haven't mastered their art, use only the cheapest ingredients, and are out primarily to make a fast buck.

Day 1966: 11

His piece said, Day goes on to tell us how to cook 'authentic' curries, though these often bear the stamp of colonial interests, with recipes for dishes like corned-beef bhurta and the common use of ingredients such as Worcestershire sauce. He also offers further commentary on curry, and pieces of helpful advice (on page 68, on the preparation of poppadams, he writes, 'They may be fried in fat, have a little butter spread on them and grilled, or grilled without any fat whatsoever. Women who have a tendency to put on weight should adopt this last method'). Comments made by Narayan on the colonial fabrication of curry, and by Appadurai on curry as the mythic master trope of Indian cuisine, are clearly supported here, in a book written by a white Englishman on how to cook 'proper' curries. The more scholarly approach of food historian David Burton has produced a rather different book (his *The Raj at Table*, 1993), but one which gives us further insight into this colonial culinary complex. Adaptations similar to corned-beef bhurta, such as Sri Lankan Christmas cake, Sandhurst curry and recipes with crab and lobster (unknown in precolonial Indian cooking) are catalogued in abundance. Even Mrs Beeton makes a guest appearance, offering a chicken curry with chickpeas. Here colonial India is a culinary curio, and recipes are provided alongside historical cameos mapping the changing attitudes of the colonisers towards the colonised cuisine.

AUTHENTICITY AND TRADITION

Our final taste of India in this chapter comes from a recent advertisement from the in-house magazine of the supermarket giant Sainsbury's, announcing its Indian Ready Meals (Plate 7.2). Its headline, 'Made by Indians, not cowboys' puns on the word cowboy meaning a charlatan who does not do a proper job (more usually applied to 'cowboy builders'), leading us neatly into a discourse of authenticity – something remarked upon by many commentators as an especially important element in discerning food choices. As the advert says,

Made by Indians, not cowboys.

It takes a special kind of person to make an authentic Indian meal.

An authentic Indian.

That's why, at Sainsbury's, we didn't ask any Tom, Dick or Harry to make our Indian Ready Meals.

We asked Akbar, Nizar and Zeenat.

People who know their poppadoms from their cardamoms.

Their tamarind from their turmeric.

And their fenugreek from their jaggery.

And if all their ingredients are authentic, so are their cooking methods.

They have a giant oven that recreates the unique kind of heat produced by a traditional tandoor oven.

Huge pans that simulate the action of tipping a saucepan back and forth to ensure all sides of the meat are exposed to the spices.

The result is Indian food that could have been made in Hyderabad, Jaipur or Bombay.

The dish making your mouth water on the opposite page is our Traditional Chicken Tikka Curry.

We can also tempt you with such delicacies as Tandoori Chicken Sizzler, Creamy Prawn Masala and Chicken Balti.

In all, there are 19 Sainsbury's Indian Ready Meals to choose from.

All the hard work's been done. The only thing you have to do is heat, and serve.

Why not pop out for a take-away today?

SAINSBURY'S Where good food costs less.

Plate 7.2 The taste of 'authentic' India?
Source: *Sainsbury's Magazine*, 1995

It takes a special kind of person to make an authentic Indian meal. An authentic Indian. That's why, at Sainsbury's, we didn't ask any Tom, Dick or Harry to make our Indian Ready Meals. We asked Akbar, Nizar and Zeenat. People who know their poppadoms from their cardamoms. Their tamarind from their turmeric. And their fenugreek from their jaggery.

The result of such care for authenticity, the advert concludes, is 'Indian food that could have been made in Hyderabad, Jaipur or Bombay'; illustrated is the Traditional Chicken Tikka Curry, sumptuously photographed in the midst of authenticating ingredients and cooking equipment. Tradition, like authenticity, is, as Emiko Ohnuki-Tierney (1993) remarks, a powerful 'invented' discourse in the presentation and representation of food and national or local cultures, often used oxymoronically by producers, as in the notice in a local pie shop advertising 'new traditional pasties'. And it is precisely in opposition to the now crass colonial incorporation and standardisation of generic 'curry' or dishes like corned-beef bhurta that 'authentic' and 'traditional' dishes such as Sainsbury's Indian Ready Meals are situated, in the spirit of multiculturalist gastronomy.

However, the problematics of thinking about ethnic food restaurants, as Narayan does above, or of the naked colonial history revealed in Burton's study, are as nothing compared with the complexities of a Sainsbury's ready meal, where 'authentic Indians' produce 'traditional' dishes for a huge retail chain which is dominated by white Western workers and bosses (not to mention shoppers). The final lines of the advert hark back to colonial exploitation: 'All the hard work's been done. The only thing you have to do is heat, and serve.' In an instant (almost), a Sainsbury's shopper can feast on a cuisine centuries in the making, and reap the benefits of the history of colonialism which brought 'authentic Indians' to Britain and 'authentic Indian meals' to the British table.

In a sense, of course, the Sainsbury's advert does try to make us think clearly about who produces the product in the way Narayan proposes – we know that Akbar, Nizar and Zeenat (metaphorically) do. What we don't know is how much they stand to gain out of the enterprise, however. Given the weight of the history of British incorporation of Indian food and Asian people, it would be difficult to describe a Sainsbury's Indian Ready Meal as 'fair trade' or 'ethical eating' – it is probably less fair to eat one than to get a similar dish from an Asian-owned restaurant or take-away, for example. As David Harvey (1989: 300) says, in the age of time–space compression, different worlds are brought together 'in such a way as to conceal almost perfectly any trace of origin, of the labour processes that produced them, or of the social relations implicated in their production'.

The discourses of authenticity and tradition so central to the advertising campaign discussed above crop up time and again in food marketing discussions. George Hughes' (1995) paper on two efforts to market Scottish food – the Taste of Scotland campaign in the 1970s and the introduction of a Scottish food quality mark in the early 1990s – shows how these discourses were mobilised instrumentally to capture a bigger slice of, in the first instance, the tourist trade, and, in the second, the more general, discerning consumer. Combining a rich cultural heritage with the environmental stereotypes of rural Scotland as an agricultural heartland, both campaigns relied almost exclusively on a territorial stamp of authenticity – Scottishness – to construct a 'food heritage' and a 'national larder' (Hughes 1995: 787) used to promote gastronomic tourism (whether combined with actual tourism or not). As consumer psychology research shows, the use of certain 'geographical product descriptors' enhances market position through such positive associations (Hodgson and Bruhn 1993).

FOOD–NATION IDENTIFICATIONS

But what can you do in a nation without a clearly marketable culinary heritage? That is the question which has faced Australia: as Michael Symons, food author, says, 'Australia has very few dishes of local origin. About the only creation for which we can claim even modest culinary significance is the pavlova' – and even that is disputable, with its origins traced back to New Zealand (Symons 1983: 122). (Of course, he is talking about white Australians here; no mention is made of Aboriginal cuisine, other than to note that, as hunter-gatherers, they have no sedentary 'agricultural' or gastronomic heritage which could be exploited by settlers.) Symons describes what he calls a 'tyranny of transport' through Australian history, which has given its cuisine a uniquely 'portable' character, without any notable regional distinctions. Developments in transport – first ships, then railways, then automobiles – define for Symons the phases of Australia's food history (respectively leading to, in Symons' (1983: 125) words, the 'industrialisation of the garden', the 'industrialisation of the pantry', and the 'industrialisation of the kitchen'). Yet Australian food remained, at least until the 1970s, definitively 'English', following English customs even when they ill-matched local conditions (such as having full Sunday roasts year-round, and a 'traditional' Christmas dinner in mid-summer) – although American culinary imperialism reached Australia in the Second World War, establishing world brands like Coca-Cola (which recently ran a TV commercial in the UK featuring Aboriginals). Searching for something to call a national dish, Symons notes (with some humour) that the prime candidate could well be meat pie and tomato sauce – emblematic of the 'artificial and careless' nature of Australian cuisine. Lupton (1996: 26) offers, as an alternative 'symbol of Australian citizenship', the jar of Vegemite; we could add the omnipresent barbecue, or 'barbie', much beloved of Australian soap opera characters, or Foster's lager, marketed in Britain as quintessentially Australian (meaning gruff, chauvinistic, macho). The equally gruff and macho alternative, 'bush tucker', has yet to catch on as anything more than a culinary curio, while the few Australian-themed eateries in the UK, such as Manchester's Tucker, serve a mix of Aussified fast-food staples – 'chook legs' or the 'stockman's lot' beefburger, for example – with the occasional exotic addition such as crocodile fillets or kangaroo steaks.

Of course, Australia has also received other immigrants, notably from Asia but also from parts of Europe, who have introduced 'foreign' elements into Australia's 'one continuous picnic' (Symons' term) – interviews in Deborah Lupton's (1996) study reveal a more diverse fare including 'Chinese', 'Thai' and 'Indian' food as well as 'Italian', among others. It has also inevitably been infected by the globalisation of foodie-ism, which is replacing the Australian table described by Symons (if only in the choicest of big-city restaurants). As food critic John Lanchester (1995b: 38) observes, the antipodes have a 'newly acquired and passionately upheld reputation for exciting cooking' (his review of two Australian chefs now cooking *in England* shows an absorption of South-East Asian, Middle Eastern and Mediterranean influences – and no meat pies, Vegemite or kangaroo in sight).

In stark contrast to the discernible lack of a national dish in Australia, some nations are frequently associated (by themselves and/or by others) very strongly with a single food item. Sometimes this is a piece of bad stereotyping – the use of curry as a trope of Indian cuisine and culture, discussed above, for example, or the advice of multinational-management consultant Martin Gannon (1994) that French wine and Turkish coffee-houses represent suitable cultural metaphors for learning how to do business with those countries. And drinking Guinness is obviously a strong connotation of Irishness, because sales of the stout rocketed as the Eire soccer team played well in the 1994 World Cup (so much so that the launch of Guinness's draught bitter in a can, Enigma, had to be delayed because all production was focused on making enough stout to meet demand). A recent British TV commercial for a rival Irish beer, Caffreys, shows a young Irish-American man transported from a US cityscape to rural Eire at his first sip of ale, continuing the association (being increasingly capitalised on in the UK by a rash of pubs getting revamped as 'Irish theme bars'). And in a bewildering turnaround of the old British tradition of visiting overseas holiday resorts and demanding English grub and beer, one British tour company now offers travel-shy gastronomic tourists the possibility of savouring the flavours of other countries without the bore of having to get there or mix with actual foreigners: its German weekend (highlight, a Bavarian banquet with all the trimmings), held in the English Midlands town of Walsall, was billed as a 'Eurosceptics tour' (Born 1995).

One of the most thoroughly researched of these food–nation identifications – and one which is thus worth exploring here – is the link between rice and Japan, written up beautifully by Emiko Ohnuki-Tierney in *Rice as Self* (1993). While other writers have explored Japanese culinary culture from, among others, a semiotic perspective (e.g.

Loveday and Chiba 1985), none has delved so deeply into the symbolism of a single food item. Rice, as the 'staple' of Japanese diet, has a special centrality in national identity formation. As Peter Bishop (1991: 32) notes, the 'core complex-carbohydrate' is often the most celebrated item on a nation's table, and also the most fundamental to a meal: 'Unless they have eaten some of this basic food, whether it be wheat, oats, barley, rice, potatoes, people traditionally do not feel that they have eaten.' Or, as Ohnuki-Tierney (1993: 41–2) puts it:

> Most Japanese continue to associate the [main] evening meal at home with rice; an evening meal without it would be equivalent to sandwiches for dinner for many Americans.... Japanese, especially older Japanese who travel abroad, often complain that they do not feel satisfied after eating meals without rice; *manpukukan* (the full-stomach feeling) is not achieved without rice, no matter what else is eaten.

And while the range of starchy staples available has certainly broadened in many national cuisines, they often retain their 'foreignness' and are not readily combined with traditional, national menus (roast beef served with rice, tagliatelle or noodles would be practically unthinkable for a British Sunday lunch). In Japan, the centrality of rice is expressed throughout national culture, with carbohydrate staples being used to distinguish between Japan and other nations; rice is, Ohnuki-Tierney says, the dominant symbol of the self in Japan (even though rice agriculture and rice eating have long been *quantitatively* unimportant for large segments of the population, especially with the popularisation of other side-dishes – and despite the fact that rice is itself an import from mainland Asia). In her study, Ohnuki-Tierney sets out a 'historicised anthropology' which explores all the facets of the place of rice in Japanese society and culture, from its position at the centre of traditional and contemporary cosmology (its past equation with deities and the belief that rice has a soul), its relation to space (land) and time (history), its role in commensality and as a gift, its aestheticisation as both beautiful and powerful and as a symbol of 'the good life', and its use as 'pure money' (compared with 'tainted' cash).

At times of crisis, these rhetorics combine with an overwhelming nationalism to give rice a unique and many-layered symbolism of *Japaneseness*. And although rice has now become 'secularised' in Japanese society, its symbolic potency remains. In terms of setting Japan apart from other nations and cultures, rice is similarly mobilised as the dominant trope – against Westerners' obsession with meat, for example (although the two are sometimes eclectically combined, with hamburgers served with rice squashed together in a bun shape, for example). Further, imports of rice (especially short-grained Californian rice) are dismissed as impure, chemically adulterated, *different*. There is a strong economic paradox attached to the issue of rice importation:

> Contemporary Japan is a postindustrial nation that is a far cry from a rice-paddy agrarian society. Government subsidies to farmers have been under attack from urbanites who make up the majority of the population [and who pay via taxation for the subsidies]. With an increase in the use of side dishes, the amount of rice consumed by contemporary Japanese has been drastically reduced. Yet rice continues to be a dominant metaphor of the Japanese self. Many, but by no means all, Japanese view the California rice importation as a threat to their identity and its autonomy. Urbanites, many of whom resent the farm subsidies provided by the government and who are antagonistic toward farmers, are still willing to pay for the *symbolic* value of domestic rice.
>
> Ohnuki-Tierney 1993: 111

THE NATION'S DIET

The note about farm subsidies brings us neatly on to the last aspect of the nation we want to consider here. Thus far we have largely been concerned with thinking the nation through food: Japanese rice, Australia's 'continuous picnic', Indian curry. Now we want to change direction to signal the ways in which the nation's diet is organised, observed and regulated by agencies of the nation-state. To do this, we want to look briefly at a number of cases of state involvement in the national diet.

Britain's National Food Survey (NFS), which celebrated its half-century in 1990, is an ongoing data collection exercise run by the Ministry of Agriculture, Fisheries and Food (MAFF), which each year receives seven-day food-diary information from approximately 7,000 households (for details, see Slater 1991). Elaborate nutritional and economic information and forecasting, such as demand analysis, is then generated from the data (Fine, Heasman and Wright (1996) wring data on food groups such as meat and dairy produce, and on socio-economic indices such as food and class or the influence of children on food purchases from the NFS). Food policy – and its critique – is obviously dependent upon survey data like these (see, for example, Frank, Fallows and Wheelock 1984; Lang 1986/7), while the discipline of nutritional science had its origins in the same health and dietary concerns which prompted the commencement of the NFS (Lupton 1996). The agenda of the NFS thus mixes state paternalism with market analysis – sometimes an uncomfortable mix.

Often the findings of food surveys prove controversial. Early in 1996 a political storm broke in the UK over the so-called 'bad food trap'. As Judy Jones (1996: 13) wrote, with typical journalese emphasis: 'You've got no *money*. You *eat* badly, and your *children* eat badly. Your family gets *sick* quickly. The doctors call this *malnutrition*. You are not alone; this is happening to *millions* of people in *Britain* today. They are caught in THE BAD FOOD TRAP'. (Extensive earlier sociological research (e.g. Charles and Kerr 1986b) had already highlighted the problems of low-income eating, but no one had seemed to notice.) Revelations about the diets of Britain's low-income families, and the health implications of consistently poor food intake, prompted media-hungry attempts to prove (or disprove) that an 'average family' could eat well enough on State benefits. (A similar incident over student grants led to at least one university vice chancellor trying to live on undergraduate fare for a week, to prove a point.)

Food surveys of various types, and with variable integrity, are a continuous feature of contemporary state interventions in diet and health surveillance and promotion schemes. In the hands of a self-confessed non-interventionist government acutely aware of the precarious balancing act between the market and social welfare, scandals like the 'bad food trap' and fears of a 'fat future' for Britain (Jones 1996) never fail to provide dilemmas: freedom to eat (and, more importantly, to buy) versus meeting future medical requirements (and the tax burdens they will bring). In other parts of the world (and in our own part, in different times), however, adamantly interventionist strategies are deployed in the face of national dietary anxiety or catastrophe. In Tonga, for example, after a survey in the 1980s revealed over half the adult population to be obese by World Health Organization standards, a national slimming competition and weight awareness campaign was instigated, and the ruler, King Taufu'ahau Tupou IV, himself took the lead, losing 11 stone (about 150 lb) and appearing on TV in the gym (Broadbent 1995). And in 1880s America, the (health and economic) dangers of a new product – margarine – were seen as so threatening that prohibition laws were passed in many states (possession of margarine with intent to sell carried a maximum fine of $1,000 or a year in jail). In places where the substance was not banned outright, restrictive laws were passed, such as those which decreed that margarine must be coloured pink. Many of the restrictive laws were not fully repealed until the 1950s (Ball and Lilly 1982). In Britain, too, margarine initially was viewed with suspicion, and only legally sold through special shops, so that it could not be passed off as butter (Fraser 1981).

Although margarine is now more or less accepted as a legitimate foodstuff, public concerns over food safety continue to prompt government action, which is not always appropriately responsive to consumer fears and behaviours (Schafer *et al.* 1993) – especially when, paradoxically, it is also reported that the 'public' is sick of being told what to eat to be healthy, approaching government dietary advice with mounting scepticism (Willetts and Keane 1995). Fears of contamination and adulteration, right up to contemporary anxieties about BSE in British beef, thus provide the nation-state with complex dilemmas. It is important to note, for instance, that an almost total rejection of beef from British dinner tables was reversed by the simple process of slashing the price of the meat on supermarket shelves, showing how the market's power can overcome a serious health scare like BSE, as well as revealing how fickle the public can be over food safety (it is similarly interesting that beef consumption outside the UK, regardless of the presence or absence of BSE in herds, has dramatically declined – domestic beef sales dropping by 65 per cent in Germany by May 1996 (Naughton 1996)). Whether or not the BSE scare is more about 'bloody stupid Europeans' than bovine spongiform encephalopathy, its emergence as a global issue in mid-1996 sits at the very nexus of food and nation which we have been exploring in this chapter.

NOTES

1 A short piece by John Naughton in British broadsheet newspaper the *Observer* in May 1996 contains some interesting twists and turns on the BSE issue: for Europhobics, he quotes tales of Germans fearful of sitting on leather sofas; for Yank-haters, he cites US Customs' refusal to allow Damien Hirst's latest pickled cow-corpse artwork through on the grounds that it contravened the worldwide ban on British beef import; but he also chides British ministers and consumers for *laissez-faire* myopia.

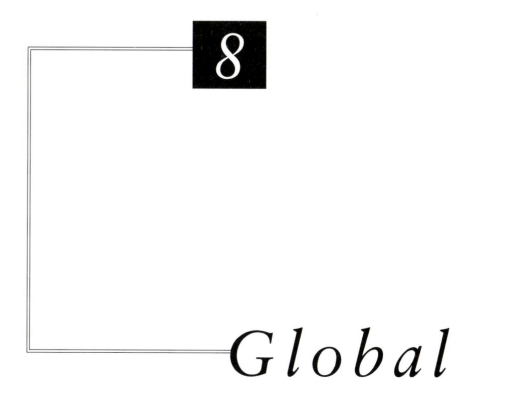

8

Global

WILBUR ZELINSKY (41)

•

I have invented and perfected a thoroughly delectable soup, something indulged in only by family and closest friends. Experimentation has demonstrated that it cures, or at least alleviates the symptoms of, whatever ails man or beast. As for taste, it makes strong men weep, and prompts the merely able-bodied to utter moans of ecstasy.

Wild unicorns could not drag the recipe out of me, and it is likely to be buried with me. All I can say is that copious quantities of lima beans are called for . . .

GLOBAL

Globalisation has become a key social-scientific concept as we near the end of the twentieth century. In a constantly changing raft of theoretical perspectives, we find those pessimistic about the future (when the whole world will resemble an American shopping mall) and those more optimistic (who see elements of distinct cultures from around the world mixing in creative ways). Borrowing from an assortment of these theories, we seek in this final chapter to tread a path through the maze of globalisation, nibbling the occasional bit of fruit (or whatever takes your fancy) along the way.

Fruit, in fact, is a peculiarly good food item to think global with, since there has been a concerted effort to bring increasingly obscure and exotic fruits from increasingly remote parts of the world, and to lay them all out alluringly on supermarket shelves. This has provoked, in some quarters at least, a frenzy of the exotic, and a fascination with authentic recipes and ingredients – part of being truly cosmopolitan, truly a citizen of the world, is knowing your rambutans from your kiwano (and knowing what to do with them).

Being a citizen of the world also means, for some people, eating only your share, and eating only what is ethical. From hippie cookbooks to macrobiotics, green cuisine has been (and continues to be) an important countercultural response to being in the world.

One thing that is especially interesting about the scale of the global used here is its relation to the other scales discussed in the book. So, we end this chapter with a set of observations about how the body and the global interact, and likewise with the home, community, city, region and nation. This offers us one more chance to review the important themes of this book, and to stress that, while the scale device has been handy for us to write around, we must always see the interconnections, which are much more numerous and complex than we can possibly show here.

8

GLOBAL

•

To begin, a selection of recent advertising slogans, cookbook titles, newspaper cookery-page headlines and other food media representations of the global:

Discover the World – and Eat It

The Global Kitchen

Around the World in 80 Dishes

Eat Your Way around the Globe

World in a Stew

The Global Bazaar

Trans-Global Expression

Eat Your Way around the World

All around the World Cookbook

World on a Plate

We could go on; the food media have truly developed an overwhelming obsession with global cuisine, seeing it as (or making it) probably the dominant trend in 1990s Western culinary culture. Of course, this fascination with globalisation is not limited to cooking and eating. As Ian Cook and Phil Crang (1996: 133) note, globalisation 'has become *the* social-scientific concept of the 1990s, fostering the same sort of booming publishing economy that postmodernism did in the 1980s'. But if we dip into any of these countless texts, we see that food is constantly used as an example of the complex processes of globalisation being theorised – it always seems to provide a perfect, commonsense, everyday way of illustrating sometimes difficult ideas. Cultural theorists are thus to be found pondering a bunch of grapes (Harvey 1990) or different cans of cola (Friedman 1990) or pancakes (Hannerz 1996) or doughnuts (Brady 1996) or McDonald's (virtually everyone): food is obviously good to 'think global' with.

COOK GLOBAL, EAT LOCAL

The kinds of ways this thinking has been done have tended, at least until recently, either to stress forces of homogenisation – of the global spread of an Americanised fast-food culture (McDonaldisation, Coca-Colonialism) – or to examine the counter-trend towards the reinstatement of the local as resistance to the homogenising global – a difference versus sameness contest. This discussion has latterly been refined to stress the complementarity of the two processes, forging a more dialectical relationship between the global and the local, and emphasising processes like hybridisation, indigenisation, or creolisation (see the essays in Featherstone (1990) and Featherstone, Lash and Robertson (1995)). According to these theories, homogenisation is impossible since global sameness is always subject to local reworking and contextualisation. So, for example, McDonald's is a very different place, with very different meanings, in Moscow than it is in Manchester or Michigan (Smart 1994); Coca-Cola similarly takes on local cultural significance in the Congo (Friedman 1990) or in a 'Third World' village (Weyland 1993). The realisation of these complex two-way processes has led to intense theoretical refinement, giving us a stress on 'globalisations' in the plural and on the 'globalisation of diversity' (Nederveen Pieterse 1995), even leading Roland Robertson (1995) to coin the term 'glocalization' as a shorthand way of suggesting inclusivity and symbiosis within the homogeneity–heterogeneity mix (and leading to corporate-strategic reorientation, hence the Coca-Cola slogan 'We are not multi-national, we are multi-local'). Of course, it is not just that global sameness is reworked by local difference – local difference is also globalised as consumers search out the 'authentic' 'exotic' to pile on their plates.

The selling (and buying) of this authentic exotic has understandably attracted some attention from geographers, since it has a lot to do with space and place (notably Cook 1994, 1995; Cook and Crang 1996; Crang 1996; May 1993). We will need to look at their work here, interspersing it both with insights from other work on globalisation and with examples of particular cases, for its role in thinking the global through food must be placed centre-stage in our discussion; but we shall also range around and beyond it to consider aspects of the globalisation of food consumption not addressed at depth in their work, and to talk about some interesting recent trends in food consumption which in some cases signal a reorientation of the global–local nexus.

To start that, here is Malcolm Waters (1995: 3), from his pocket-book summary of globalisation, defining it as 'a social process in which the constraints of geography on social and cultural arrangements recede and in which people become increasingly aware that they are receding'. This is an interesting definition, partly because it instantly raises a lot of questions: who is 'aware' of these processes? Do the constraints of geography recede for everyone? (see Grossberg 1996 for criticism of globalisation theories along these (and other) lines). Waters is both spatially and temporally unspecific here – yet lots of the theoretical work around globalisation has been concerned with precisely such specificities. Obviously, in some senses globalisation is a very old process; writers such as Harvey (1989) make this point, discussing the historical sequences of transportation and communications advances, for example, which have effectively made the world 'smaller' for lots of people (a process he names 'time–space compression'). The importance of colonialism must also be noted, since much of what happens in the world today – including our knowledge about it, and the ways we use that knowledge in consumption practices – bears the mark of colonial activities and legacies (as histories of tea or sugar, for example, make apparent; see Hall 1991; Mintz 1985; Smith 1992). What is also clear, however, is that the process is accelerating at the end of the twentieth century, driven by corporate expansionism, media and telecommunications 'revolutions', changing consumer consciousness, and so on.

Most commentators in fact stress consumer consciousness – or reflexivity – as key to understanding globalisation. As Waters' remark points out, it is our knowledge that the world is getting smaller which is crucial, and the uses we make of that knowledge: our increasing understanding of the world and our place within it drives globalisation. This has led to work exploring the roles of travel and tourism – as knowledge-gathering practices – in our everyday lives (Hannerz 1990; Urry 1990), extended to include the 'touristic quality' of much of contemporary consumption, where cultural 'snapshots' can be experienced, gazed upon and collected in a stroll round the supermarket or lunch at a restaurant (Lash and Urry 1994; Smart 1994), and where 'kitchen table tourism' is facilitated by the provision not only of foodstuffs but also recipes and inspirations from the proliferating food media (Curtis and Pajaczkowska 1994). This notion is one also commonly employed in the marketing strategies used to sell us things (the *Time Out* 'world on a plate' example discussed in Chapter 5 is a clear illustration of this). Other groups of people in movement – migrants, refugees, guest-workers, exiles – also contribute to reshaping the global cultural (including culinary-cultural) landscape, as do other world-spanning cultural flows, including those of technologies, media, finance and ideologies – flows which criss-cross the globe in 'disjunctive' ways, creating the 'uncertain landscapes' or 'imagined worlds' symptomatic of contemporary globalisation (Appadurai 1990). Commodities have a powerful role in these imagined worlds: as bearers of many of the symbols of globalisation, they are routinely used to articulate both place and movement – and, through those, identity and identification.

This combination of place and motion leads Cook and Crang (1996) to use the notion of 'displacement' to think through food's global cultural geographies. This is a more useful way of considering globalisation's impacts on consumption than those of homogenisation versus indigenisation or creolisation, they argue, since it emphasises that there are no 'pure cultures to mix', and also that the processes are not 'some recent disturbances of past cultural forms' (Cook and Crang 1996: 139) – stressing the longevity of globalisation. As has been noted in earlier chapters, culinary cultures constructed as 'original', 'authentic' and place-bound – regional or national cuisines, for example – can be deconstructed as mere moments in ongoing processes of incorporation, reworking and redefinition: food is always on the move, and always has been. As Crang (1996: 47) spells out in another paper on displacement and consumption:

> In broad terms the figure of displacement is used here to suggest an under-standing whereby: processes of consumption are cast as local, in the sense of contextual; but where those contexts are recognised as being opened up by and constituted through connections into any number of networks, networks which extend beyond delimiting boundaries of particular places; where imagined and performed representations about 'origins', 'destinations', and forms of 'travel' surround the various flows of people, goods, and services in these networks; and where consumers (and other actors in commodity systems) find themselves positioned and position themselves in terms of their entanglements with these flows and representations.

Central to this is the creation of geographical knowledges or 'lores' about commodities, which are produced by the actors in commodity systems – producers, retailers, promoters, and consumers. This may invoke what Cook and Crang (1996: 132) describe as a 'double commodity fetish', incorporating certain *ignorances* – about production conditions, for example (as noted by Harvey 1989) – together with certain place constructions or *knowledges*, such as those about 'origins' (especially those which stress naturalness, or exoticness, for instance). As Ferguson and Zukin (1995: 197) put it succinctly, '[t]he study of food is, in good part, a study of the culinary "worlds" produced by the many interactions that put food on a table, ready to eat'. Cook and Crang (1996: 141) refer to these worlds as 'circuits of culinary culture'.

IT'S A QUEER WORLD

Perhaps it is time to think about a particular example. Probably the best-known and most-used commodity for thinking the scale of the global through food is exotic fruit –

known in the British trade, together with exotic vegetables, as 'Queer Gear' (Heal and Allsop 1986). The rise of fresh exotic fruit and veg consumption in the UK – and particularly the role of supermarkets in selling the stuff and the food media in encouraging us to try it – has been a remarkable cultural phenomenon (Cook 1994, 1995). According to one market research report, exotic vegetables rank behind only pickles, curry powder and 'ethnic snacks' (such as poppadams and prawn crackers) in the UK's 'ethnic food' popularity charts, with the largest group of purchasers in social class AB and the 25–34 age group (Keynote 1995).

The rise of exotic fruit is, of course, also the continuation of a long process of bringing produce from overseas to the British table; much of the fruit which we now think of as definitively unexotic – the banana, for example – was once viewed in the same way we now gaze upon the kumquat or the grenadillo. In her extraordinary *A Primer for Daily Life* (1991), Susan Willis makes this point in an exploration of banana symbolism in William Faulkner's 1920s novel *As I Lay Dying*; a similar point is made by De Vita (1994) in relation to pizza's de-exoticization in the USA. As Ian Cook (1994) notes, other fruits, such as mangoes, avocados and kiwi-fruit, have more recently made the transition from 'exotic' to 'everyday', at least in supermarkets' eyes (see also Henderson 1992). This process of the trickling down of taste, bringing the kiwi fruit into many more kitchens, means that the exotic produce sector is constantly searching out new recruits, sourcing new exotics to tempt discerning shoppers. Important here is what Cook (1994: 232) calls the 'symbolic production' of the meanings of a fruit or vegetable – and the meaning and identity it thus confers on those who buy it.

The search for the 'exotic' in food, often related to the quest for the 'authentic', is usually seen as the territory of the so-called 'new middle class' or 'new service class' (many of whom might once have been called 'yuppies') – as suggested by the Keynote survey cited above. Interviews conducted by Jon May (1993) with new service class members reveal much about their opinions regarding 'authentic' and 'exotic' food:

> 'I love it because it's different ... a little taste of something more exotic, like lemon grass in soup.... Most days ... I might have an Indian meal, or a Chinese meal, or a vegetarian take-away, or pasta, um.... I never go in and have a cheese omelette – never, it's just boring.... It's lovely having the opportunity...'
>
> 'Ruth', quoted in May 1993: 5

What May's respondents show is a sophisticated knowledge about the foods they eat (one talks at length about cappuccino, moaning about its corruption from the authentic Italian to the pale English imitation), and pride in that knowledge. Here we revisit Bourdieu's work on distinction and cultural capital: part of the marking of difference is via knowledge used in consumption practices, such that buying exotic fruit marks the consumer in certain ways (as sophisticated, educated, adventurous, cosmopolitan, worldly wise). As Deborah Lupton (1996: 126) writes,

> It could be asserted that in the context of western societies at the end of the twentieth century, diversity in food choice is considered more important, and is available to more people, than ever before. Differentiation and innovation are highly culturally valued. In the context of an abundance of food, the search for new taste sensations and eating experiences is considered a means of improving oneself, adding 'value' and a sense of excitement to life.

Some interesting things have happened recently in the presentation of exotic produce for British consumers, however. First, where the focus of attention has previously been on fruit and vegetables as emblems of the exotic, interest is now growing in meat and fish products, a trend that has been given an added boost by the beef crisis. Kangaroo, emu, ostrich, crocodile, shark, snake, parrotfish, snapper, bream, zander and assorted other animals have recently begun to appear in British shops and restaurants (Durham 1996; Tredre 1996). And second, there has been, in some camps, a foodie withdrawal from the global search for exoticness, and a turn towards stressing the 'exotic within'; this includes organic specialist the Henry Doubleday Research Association's Adopt-a-Vegetable scheme (allowing patrons to pay towards safeguarding varieties of British fruit and veg, many of which cannot legally be grown because of EU regulations), a growing interest in offal and a fascination for 'traditional British grub' like fish and chips (Nigel Slater (1995a: 19) has declared 1996 'the year of the gastro-yob'), and the appearance of food-media stars actually *criticising* the fetish of the exotic:

> In their own climate and eaten at the point of perfect ripeness, such 'queer gear' (as it is charmingly known in the trade) is the stuff of dreams and some of it travels very well indeed. But the stores seem to be missing the point. Why race to find the weirdest fruit in the world when it wasn't worth finding in the first place?
>
> Slater 1995b: 42

Such signs that world-weariness might be replacing the worldly-wise attitude associated with fondness for exotic foods must be worrying for supermarkets which have invested heavily in establishing commodity chains based upon consumer desires for the weird and alien. They might also be taken – together with other consumption trends such as ethical eating and green consumerism – as signifying a shift in the politics of food consumption. While exotic fruits might be eaten in a spirit of multiculturalism, they are now causing distaste in some mouths on (at least) three counts. First, there is discomfort around the production realities, which are periodically exposed – the recent disclosure of the 'dollar banana belt', where workers are falling ill from over-use of agrochemicals, and where the environmental impact of over-producing banana monoculture is wreaking havoc, being but the latest example. Second, there is unease about the role export pressure of exotic produce plays in sustaining and even deepening inequities in new global realms of capital accumulation dominated by transnational corporations (TNCs) (as McMichael and Myhre (1991: 94) note, '[t]he new export-oriented luxury food agro-industry is the fastest growing sector [in the global food regime], accounting in 1980 for 25 per cent of the Third World's total processed food output, much of which is marketed by a handful of transnational corporations'). Third, there are qualms about excess in Western consumption generally (increased if, as Slater suggests, the stuff isn't even worth eating). Taken together, these factors might be contributing to the turn away from the exotic and towards an assortment of globally aware alternatives.

DIETS FOR A SMALL PLANET

Since the publication of Frances Moore Lappé's *Diet for a Small Planet* in 1971 – if not before – there has been an explosion on a par with that in globalisation texts of books detailing the 'world food problem'. Susan George's *Food for Beginners* (1982) and Geoff Tansey and Tony Worsley's *The Food System: A Guide* (1995) are representative, and both outline the issues involved in how the world gets (or doesn't get) fed. Books like these have become commonplace on Western shelves, often sitting uneasily next to 'round the world' cookbooks and well-thumbed copies of the latest Delia Smith or Martha Stewart gastro-pornfest. The tales told, the knowledges provided, are familiar ones: the industrialisation of food production, the continuing emphasis on livestock farming, corporate greed, capitalist consumerism, colonialism, world debt – which add together to create parallel situations of overabundance and life-threatening scarcity. These are problems with long histories and a lot of power and money invested in them:

> The modern food system is not inevitable but has deep historical roots which are bound up with humankind's various attempts to control the biological, socio-economic and cultural aspects of food. The interplay of the forces involved has shaped the food system, producing food shortages and surpluses, hunger and overnutrition, technological brilliance and junk foods in the same world.
>
> Tansey and Worsley 1995: 24

For Western consumers, the knowledge that the world's food is in such a mess provokes many different responses. The first is charity – from donating a few old clothes to Oxfam to going to the Live Aid concert and buying the latest Comic Relief video (the latter responses mesh much more comfortably with late-modern consumer culture, making their appeal – and success – much greater than a town centre collecting-tin shake). The second is 'ethical consumerism'. Eating a diet for this small planet means, for many people, conscientious or responsible dining (Heldke 1992), making choices about products which take account of issues like 'fair trade', aid, exploitation of developing countries, environmental impacts, workers' rights and so on (Gabriel and Lang 1995), as Jo explains in Box 8.1. Warren Belasco's (1993) account of the American 'counter cuisine', discussed in Chapter 4, contains much detail on this issue. Brands like Cafédirect thus market 'ethical coffee', revealing rather than masking the links in the commodity chain – often personifying them by showing us those who will benefit (directly) from our purchase (see Plate 8.1). 'Eating green' – whether that means simply using fewer processed foods, avoiding additives, going organic, eating some construction of a Mediterranean diet, or the whole-hearted adoption of Zen macrobiotics – is an allied consumption response which often arises from similarly global concerns.

Box 8.1 JO

JO WORKS IN A SHOP WHICH SELLS FAIRLY TRADED GOODS AND ORGANIC VEGETABLES. SHE IS VEGETARIAN AND IS INTERESTED IN FOOD POLITICS AND FOOD ETHICS.

'I prefer not to use Safeway where possible; I prefer to, you know, get the things like rice and beans and lentils and fruit, fruit and veg from a local organic veg shop, and then any other things I prefer to get from the smaller, the other smaller shops like the newsagent, which has got a small range of groceries, or the Co-op, because I prefer to support smaller businesses rather than the great big multinationals.

'I think that people in Western countries consume far more, well, I mean, they're a million times over their fair share, and I think that you know, more has got to be done about getting people to eat less meat because then the land that's used to feed up the cows or whatever can be used to grow grain to feed the people in their own countries rather than exporting it over here.

'I didn't buy South African goods when, you know, before apartheid was abolished, but then my mother didn't either so it was like I grew up with that. I won't buy Nestlé products but, I mean I never drank Nescafé anyway, I always buy the cheapest coffee I can find, or the near cheapest, but I have stopped eating Nestlé chocolate bars. So anything that's produced by Nestlé I won't eat. Won't buy anyway. I might eat it if another person has bought it and I don't know, like if someone offers me a cup of coffee I will drink it, 'cos they've offered me it.'

Plate 8.1 Cafédirect: drinking coffee improves lives?
Source: *Sainsbury's Magazine*, 1995

Boycotts represent a further consumer strategy – not buying 'dollar bananas' being recently advocated in the UK and USA, to stay with our earlier example; while for many vegetarians, refusing meat is a response to the contribution made by over-intensive livestock farming to the world food problem (Beardsworth and Keil 1992). Well-known campaigns, such as those against McDonald's (accused of encouraging deforestation, unsustainable farming and countless other environmental menaces) or Nestlé (for its role in marketing infant formula in developing countries), focus activism while shedding light on the often shoddy ethical positions and policies of TNCs more broadly. And it should be remarked that a successful boycott of a single product in the West can make companies rethink their policies elsewhere (in some cases – though in many they have merely 'repositioned' products, by describing them and marketing them in different ways (Goodman and Redclift 1991) or have just cynically 'cooked the books' by creating yet more ignorances and silencing dissent, as in the 'McLibel' case).

Returning to the question of knowledges introduced above, it is interesting to note that, just as the consumption of the 'authentic' 'exotic' is in part an exercise of knowledge – and remembering that that is part of the pleasure – dining in the knowledge that you are minimising the effects of your feast on the world is also seen by many as a source of pleasure. Two respondents in Deborah Lupton's study made this exact point: Patricia talked about 'cooking with respect for the ingredients you're using', and Simon said that '[i]f you are taking your nourishment with awareness ... you can see the wonder of all those processes that led to you taking that nourishment to fuel your body' (both quoted in Lupton 1996: 88). She also cites Wendell Berry (1992: 378), who writes, 'A significant part of the pleasure of eating is one's accurate consciousness of the lives and the world from which food comes.' These knowledges inhabit exactly the spaces of the ignorances described in Cook and Crang's 'double commodity fetish' of global produce. As Lisa Heldke (1992: 301) puts it:

> I am interested in considering how my ordinary, daily activities involving food – buying, preparing, and eating it – connect me with other workers in the world who grow the food we in the United States eat. How might I use my understanding of these connections to shape my actions? Despite the real interdependence that exists between U.S. consumers and farm workers – in the United States, Mexico, Costa Rica, and Kenya, for example – these connections are often conveniently obscure or invisible to middle-class American consumers, and thus do not inform our decision-making in the grocery store.

Such thoughtful practice in cooking and eating, then, is seen as an important response to the individual powerlessness many people feel when faced with TNCs and the world food problem; it is this kind of practice which companies and products like Cafédirect tap into, by providing us with the knowledge (and the produce) we need in order to eat (and drink) responsibly. It is an important part of the process whereby consumption is reoriented towards ethical positions – and if enough people make that shift, then maybe the corporations will genuinely make it too (or so the argument goes).

Crang (1996) signals a welcome note of caution over this strategy, however. While these knowledges help to 'thicken' the connections between consumers and producers (by showing us something of what has gone on to bring our food to the table), they are open to 'reintegration into ... consumptive circuits and their productivities', creating what he calls 'geographical capital' – forms of knowledge 'liable to be utilised in consumption as coinages in processes of cultural and social distinction' (Crang 1996: 57). Hence, as Belasco's (1993) sometimes depressing history of the USA's countercuisine shows, seemingly radical practices can always be recuperated and sold back to us (often with an 'ethical premium' upping the price-tag). One of Crang's alternatives to this thickening (but one he rejects) is 'an ethical refusal to find or construct meaning at all, a radical (and hard-fought) passivity that opens up spaces in which consumed objects, subjects, and arenas can just be' (57). This might be hard to imagine – to eat without thinking – but one of Lupton's interviewees, Gilbert, confessed to having what he thought of as 'a food disorder of a sort, for he does not enjoy thinking about or eating food' (Lupton 1996: 143). While this is hardly a 'radical passivity' (especially if Gilbert considers it a disorder), it is an interesting case for thinking about Crang's point – and for connecting the global right back to our first spatial scale, the body (the same, of course, can be said for 'ethical eating').

Later in this chapter we want to carry out more of these tracings through and between scales, but first we need to stay with Phil Crang's work on displacement; for if he rejects both thickening contextualisation and radical passivity, what does he suggest as a productive, progressive way of dealing with contemporary consumption? His answer is the critical deployment of displacement, via 'aesthetic reflexivity', to think about the 'juxtapositional character of arenas of consumption' (Crang 1996: 58). This would seem to involve an acknowledgement of the fabrications inherent in the construction of consumer cultures, objects and subjects (remembering they are made, as 'active social constructions'), and a turn towards seeing those fabrications as a way of reading through and beyond 'globalist gloss' to show instead the 'connections between the "different" elements, [and thereby] to refuse the equation of difference with separation' (62). This could then provoke a destabilisation of those fabrications which might otherwise bring these different elements together to create certain (dare we say hegemonic) consumer knowledges. This involves 'roughing up' the 'surfaces' of commodities which are more usually presented to us as smooth and unblemished. Juxtaposition and entanglement are Crang's key terms here – as consumers, we are entangled in all sorts of commodity geographies, and we need to think through and beyond those, emphasising each commodity's biography as it moves through these geographies. Consumers have numerous points of contact with commodity flows and networks, and while these may be cast as place-specific or 'local', our task is to view them as juxtapositional moments – snapshots from commodity biographies which can be set against other snapshots, other moments. To quote a neat example from Cook and Crang (1996: 148), this could involve 'setting supermarkets' use of paradisical discourses of tropical fecundity and simplicity next to their past legitimations of colonial conquest and exploitation' in the case of our old friend, exotic fruit – not to say one is a bad or false representation and the other good or true, but to let those images bang up against one another, perhaps to produce a 'third knowledge' beyond, between and through them.

Crang concludes that all this can only be approached through theory which combines political economy and cultural studies – but not a comfortable blend of the two; it must be a destabilised, juxtaposed, displaced theory again sitting uneasily between and beyond them. Such an approach would not get dragged into debating 'the authenticity of accuracy of commodity surfaces', but would be concerned with 'the spatial settings and social itineraries that are established through their usage' (Cook and Crang 1996: 148). As Robertson *et al.* (1994a: 6) conclude, '[p]erhaps, then, the thing to do is not to tour but to detour or *detourn*' (where 'detourn' refers to the Situationist practice of reworking existing cultural forms to create new meanings precisely through acts like juxtaposing and recontextualisation – putting political critique into the speech bubbles of cartoons, for example; see Knabb 1989).

While we might be a little hesitant about some of Crang's theorising (not least about its density), the figure of displacement, as a 'sense of geographical worlds where cultural lives and economic processes are characterised not only by the points in space where they take and make place, but also by the movements to, from and between those points' (Cook and Crang 1996: 138), is useful for thinking about food and globalisation. Let us look at another example. In 'Delicious doughnuts in Berlin: the dilemma of political community in the age of global capitalism', John Brady (1996) asks a simple question: 'What does it mean to eat doughnuts, a quintessential American snack, in a non-American setting?' (2). We could say it signals the march of Americanisation and the erasure of local/regional/national snack-foods, or ask questions about doughnut vendors' working conditions, etc. We could, adopting the radical passivity stance, say that it's just a doughnut, to be eaten without a thought. Or we could think, as Brady himself does, about how

> the doughnut can be deployed in different processes of social identity definition. The doughnut as something typically American is not bound to its American context, but rather can cross borders and take on meaning for individuals outside of a purely American context.
>
> (Brady 1996: 3)

This cosmopolitan doughnut, then, could instead be read by setting questions of America's relationship with Germany (especially the post-war love–hate history of divided Germany) alongside other outbursts of food nationalism there (the case of Turkish guest-workers and doner kebabs forming a useful contrast; see Caglar 1995), *and* alongside issues in the global mobility of cultural products, *and* alongside questions about the changing roles of snack-food in contemporary societies, *and* alongside a consideration of the increasing fluidity of points of social identification that might coalesce around a doughnut, and so on. By juxtaposing these elements, but not claiming any as more 'authentic' or 'accurate' readings of a Berlin doughnut, we can begin to see how displacement – thinking place and movement – is a useful approach (especially if accompanied by a damn fine cup of coffee).

REFRACTING SCALES THROUGH THE GLOBAL

Now we want to see how we can think about food consumption across, between and beyond the handy device of spatial scale we have deployed through this book. For it is important not to overemphasise the separateness of these scales, but to see them as all mutually constitutive and entangled. To do this we shall try to reflect and refract the spatial scales of body, home, community, urban, region and nation through that of the global; indeed, the process has already begun, at least implicitly, and many readers will surely have made some connections along the way. All we offer here are some illustrations – serving suggestions which each cook might want to adapt to suit both their larders and their talents. We realise that the interconnections and disjunctions are much more complex and multilayered than this; we leave it to you to compose your own grander recipes and menus (take, if you wish, inspiration from Wolfgang Puck!).

If we think about the global and the body, then, we have already noted some examples. Ethical eating is a clear one, since every mouthful is written through with global concerns. So, many people make choices about the food they buy and eat explicitly from the perspective of a body–global nexus; as a way of making some kind of difference to the world (however small or even illusory it may be). And while asceticism might once have been about physical denial and attendant spiritual self-fulfilment (Lupton 1996), for many consumers in contemporary societies it is about global responsibility, eschewing decadence to save the earth (a number of charities encourage such connections, running sponsored twenty-four-hour fasts, for example). At this point we might want to note two (only half-serious) countertrends recently reported. The first is self-styled 'anarchist–consumerist–terrorist organisation' Decadent Action, whose choice of activism is to sip cocktails in kitsch London bars with names like the Waikiki Lounge, in the hope of killing capitalism by giving it too much of what it wants, till it dies an 'Elvis-like' death, bloated and over-indulged (Hodgkinson 1995: 8). The second is the fad for so-called 'pleasure revenge', celebrating and indulging in the high-life of conspicuous culinary consumption in response to the oppressive clean-living ascetic moods of the 1980s. Warren Schivelbusch (1992: 228) suggests that this mood of ascetic self-denial brought products like mineral water to fashion-conscious palates, its 'neutrality' being prized as a source of 'narcissistic well-being'. Pleasure revengers scoff at such restraint, instead smoking huge cigars, drinking expensive wines and cognacs, eating big steaks and *not caring* (Sullivan 1996). A parallel trend towards excessive culinary consumption, called *da chi, da he*, has also been noted in China, where it signals a different kind of consumer politics (Fan 1994). While these might be little more than fringe activities, they all signal a new take on the body–global intersection.

As Deborah Lupton (1996: 155) stresses, the many meanings of food, including those above, all contribute to the never-ending 'project of the self' (a better term than Schivelbusch's 'narcissistic well-being', we think). The contemporary valorisation of the 'neophilic consumer' – always on the look-out for new taste sensations and eating experiences, always enacting Bourdieu's quest for distinction – means that the scale of the global is particularly important to those self-constructions. Eating one's way around the world is, then, as life-enriching a project as the grand tours undertaken by European aristocrats in previous centuries. An important component of this project, Lupton suggests, is the transgression of the edible–inedible divide; hence the craze for obscure, exotic meats.

One way to think the global and the home that might prove fruitful is by considering the impact of global technological transformations upon the domestic consumption of food. A whole history of food technology advances has impacted on home cooking and eating, from refrigerated transportation to the microwave, and from television cookery shows to supermarkets on the Internet. Many of these have facilitated the project of the self-through-food, bringing new devices to match (or make) new desires (see Hardyment 1988; on domestic technologies more generally, see Silverstone and Hirsch 1992). The phenomenal growth in 'white goods' (as kitchen technologies are called) might not have readjusted the balance of domestic labour (Cockburn and Ormrod 1993), but it has filled our homes with fantastic gadgets to cook with. And while David Morley (1991) has said that the global meets the local in the sitting-room (via television viewing), this is equally true of the kitchen or the dinner table. In fact, as home style gurus have predicted, the 'living kitchen' is more important, more central to domestic life in middle-class 1990s Britain than the increasingly redundant 'living-room' (J. Young 1994); 'kitchen table tourism' is certainly seen to be replacing 'armchair tourism' for stay-at-home travellers. Hence the disjunctive global flows of 'technoscapes' (Appadurai 1990) enter our homes, impacting on our everyday lives, including the foods we eat.

Of course, kitchen table tourism and armchair tourism are neatly linked together by the food media; cookery shows, magazines and newspaper columns bring exotic delights into our homes, and encourage us to put the world on our plates (and in our ovens) – the link is made even more explicit in hybrid forms which link travel and cookery, such as the popular British TV series *Floyd on France, Floyd on Oz* and *Floyd on Africa* (Hardyment 1995; Stephenson 1996). As cookery goddess Delia Smith put it, talking about her role in the launch of *Sainsbury's Magazine*: 'You can now go into the supermarket and shop globally ... now, seeing that that's all been made available to us, we can take it one stage further and find out what to do with these ingredients' (quoted in Stephenson 1996: 2). The food media's role is thus absolutely central – even more so, it might be argued, than that of the supermarkets which import the produce, since, as Ian Cook (1994: 232) says, 'just because they [exotic fruits and vegetables] are produced and packed in one place and shipped, ripened, and delivered fresh to a store in another, it does not necessarily follow that anyone will buy them' (in-store magazines such as that produced by Sainsbury's are thus crucial 'infotainment' resources for marketing new products). Barry Smart's (1994: 170) discussion of cookbooks as 'gastro-porn', tempting us into 'epicurean/gastronomic form[s] of *flâneurie*', should be noted here – for the seductions of these lavishly photographed and appetising texts induce the irresistibility of their creations. As Daniella Stephenson (1996: 5–6) writes, 'the food media are an inherent [we might say central] part of consumer culture, and are most used by those wishing to articulate a certain lifestyle', since they 'help to attach social meanings to certain products' – and, as she says, the foodstuffs themselves can often become 'celebrities', imbued with 'star qualities'. (See also Cook 1994 on the kiwano, or horned melon – ironically this is the very fruit which so outraged Nigel Slater (1995b) with its pointlessness and tastelessness.) The continuing advance of communications technologies provides ever more sites for 'gastro-porn', with wine-tasting on CD-ROM and virtual cookery on the worldwide web (Ehrlich 1995) bringing every gastronomic possibility into our homes.

We stay with the popular media (albeit a different branch) to begin thinking through the links between community and global. When a motley crew of US pop stars came together and sang 'We Are the World' as a response to the Ethiopian famine, they were articulating a notion of the 'global community' which we have already remarked upon (and see Howes 1990). Here those two scales are connected seamlessly, read as one, rather than being seen as worlds apart (for community, like region, is often cast as the local to counterpose the global). Some of the other forms of community we talked about earlier are also closely linked with responding to the global, in ways at once very similar to and very different from USA For Africa – the American countercuisine, for example, was a 'grassroots' organising of an activist community to resist certain global-capitalist processes and practices (Belasco 1993). This creation of a 'community of consumers' (or, in some cases, of anti-consumers) politicised around food issues, and the more focused building of communal, co-operative forms of life to address those issues, clearly sits at the intersection of the scales of community and global.

Food often gets used to articulate defensive uses of community, as resistance to globalising (or modernising, or otherwise threatening) forces, in fact. Thus, the inhabitants of the New Jersey Pine Barrens delineate an 'us' and a 'them' through their views about food, which becomes a 'badge of identity' and a powerful boundary marker (Gillespie 1984: 148). Yi-Fu Tuan's (1974) 'local patriotism' or Doreen Massey's (1993) 'introverted sense of place' are thus taken to the table, and exclusion from that table signals exclusion from the community. Even for diasporic communities – maybe especially for them – maintenance of 'traditional' foodways thus becomes a way of shoring up group identity: 'food links people across space and time, so that it helps create a bond with past members of the group as well as between living ones' (Kalcik 1984: 59); in contrast to the neophiliac fascination with 'eating the other', this means only 'eating the same' (or 'eating the self'). The commodification of community identity can thus provoke ambivalent responses, as the radical stance of the 'Proud Coonasses' on the popularity of Cajun cuisine shows (Gutierrez 1992).

When it comes to thinking about how the global and the urban interrelate and interconnect, a good place to focus on is the so-called world city. These cities are the major players on the world stage, financial and business centres, and home to the major media and culture industries (King 1990). In sum, they are, as Paul Knox (1995: 232) says, 'both cause and effect of economic and cultural globalization'. They are sites for what Knox calls 'global metropolitanism', centres of the compressed, 'fast' world, and home to the 'transnational producer-service class', for whom world cities 'not only represent their workplaces but are the proscenia for their materialistic, cosmopolitan lifestyles, the crucibles of their narratives, myths and transnational sensibilities' (Knox 1995: 243). Ulf Hannerz's (1990, 1996) work on the figure of the cosmopolitan is, of course, important here, with his or her 'willingness to engage with the Other' (Hannerz 1990: 239) feeding, among other things, reflexive neophiliac consumerism:

Some would eat cockroaches to prove the point, others need only eat escargots. Whichever is required, the principle is that the more clearly the alien culture contrasts with the culture of origin, the more at least parts of the former would even be seen with revulsion through the lens of the latter, the more conspicuously is surrender abroad a form of mastery at home.

Hannerz 1990: 240

City promotion often stresses world cities as 'cultural marketplaces' (Hannerz 1996), using food as a handy metaphor for the diversity offered up for sale, as we have already seen; the marketing of ethnic restaurants to tourists and the filling of metropolitan supermarket shelves with global produce (to feed all those cosmopolitans) are clear edible indications of the ways in which cities contribute towards globalisation, and globalisation contributes to city cultures.

Like community (and unlike the world city), the scale of the region sometimes faces the global defensively, seeing forces of globalisation as representing a threat to regional distinctiveness and specialism. The case of the Henry Doubleday Research Association stands as an example here, preserving 'native' varieties of fruit and vegetables and refusing or resisting processes of homogenisation in the fresh-produce sector – usually in the form of Golden Delicious or Granny Smith's apples put on our supermarket shelves by huge corporations like Cape or Del Monte (we will come back to apples in a moment, since they have quite an iconic status in British regional food identities). Part of this battle is over legislative restrictions from bodies such as the EU – always demonised as ruling on acceptable and unacceptable products and thereby threatening to erase regional and local specialities. But just as often the root cause is seen to be the relentless marching forward of totalising global capitalism. This is turn provokes defensive counter-strategies, whether that means offering vegetables up for adoption or trying to formalise the place-specificities of products, as in the case of wine regions (Moran 1993). Regions rarely get cast in the same mould as cities – as global marketplaces; theirs is an emphasis on the local. British campaigners Common Ground, for example, make much use of locally distinct foodstuffs and food habits in their nostalgic representations of 'authentic' regional diversity. On 21 October 1995 they ran 'Apple Day', its aim to encourage us to

> [c]elebrate a few of the 6,000 varieties of apples we can grow in Britain and the cultural richness they symbolise. Dig out local recipes, play apple games, go on an orchard walk, learn to prune and graft, identify your trees, taste a few ciders, give an apple to a friend to wish them good health.

That this advert for Apple Day appeared in, among other sites, the 'new left' journal *Soundings* gives us an indication of the likely constituency for Common Ground's appeal (see also Crouch and Matless 1996). We do not mean to suggest that the region is always something used defensively against the global; as Nigel Thrift (1990a) said, the changes in the world today are reshaping but not erasing the region, and there are plenty of productive interconnections between the two. But, as we emphasised in Chapter 6, the region is still commonly articulated as a place of tradition – local, place-specific tradition. New regional geographies are often implicated in older regional geographies, even if they relate to and articulate them in 'new' ways.

To start thinking about the nation and the global, here is something light from British newspaper the *Observer*'s satirical column 'Babylon'. Penned by someone with the *nom de plume* The Player, it recounts the dilemmas facing New York mayor Rudolph

Giuliani, who was organising a celebration dinner for delegates from the United Nations:

> When a host has every country in the world to think of (except Iraq, Cuba, Libya, Iran, North Korea, Somalia and Yugoslavia – all snubbed), and every country has its own culinary sore spot, you have to plan carefully. . . . Mr Giuliani's shopping list . . . was long on negatives: no beef, no shellfish, no pork, no alcohol in the food preparation and (for some reason) no white linen.
>
> Finally, a food that caused no offence to anyone (except for the vegetarian countries, who had their own menu); the true Esperanto among world cuisine is chicken.
>
> The Player 1995: 7

While the apparent near-global palatability of chicken might strike some of us as odd, this example shows something of the breadth of culinary diversity among the nations of the world – and the problem of finding a common denominator. In the wake of the beef crisis in Britain, newspapers carried features on (other countries') national eating habits, under the guise of presenting alternatives should BSE kill off the beef trade permanently. James Hamilton-Paterson's 'The rrruff!! guide to eating' (1996) is fairly representative, trotting out recipes for dog (hence the punny title), assorted insects, and a number of other culinary curios. There is something slightly salacious about the tone of these pieces, which can never fully escape a mix of voyeurism, revulsion and mild xenophobia.

Measuring nations by their eating habits is something routinely done, of course. For some, national diversity is something to celebrate (and even to capture through commodification), while for others it is a source of dread (one wonders if Mayor Giuliani would not have been better opting for a pot-luck feast). As globalisation turns its emphasis on diversity and difference, the culinary trend-setters head off (literally or metaphorically) into increasingly far-flung corners of the world, to seek out new treats and tastes. While this 'gastro-global eclecticism' (Smart 1994) has been criticised by some commentators for producing 'shallow' knowledges about nations and sometimes using crass stereotyping along the way (e.g. Diamond 1993), for others it is a cause for celebration – it is seen as both transnational and nation-specific (but anti-nationalistic). It is also seen as an antidote to homogenising forces of globalisation – which are never really given anything but bad press, except where they might lead to creative, contextual reworkings. Take contemporary New Zealand cuisine, for example. Here a 'national' culinary culture is being forged from a 'global sense of place', along the lines of the 'postmodern' Californian cuisine championed by the likes of Wolfgang Puck, blending 'Pacific Rim' influences with Old World favourites in a 'mix-and-macho style' (Bateman 1996: 52). Set against this reformulation of national cuisine as being about the entanglements of cultures that meet in one place, we have defensive articulations of food nationalism; Ohnuki-Tierney's (1993) brilliant *Rice as Self*, for example, shows the emblematic status of one everyday staple in the construction of national identity almost as long on negatives as Mayor Giuliani's shopping list: rice versus not-rice (Japan is not bread-eating Europe), right-rice against wrong-rice (Japan is not Korea or China).

And so we come full circle. As we said at the start of this section, we have only attempted to trace some of the interconnections and disjunctions between scales; they are obviously much more complex and co-implicated than we have sketched. It is not hard to imagine all scales meeting up on a single plate, in fact. Let that be our final thought, and we simply give an instruction (actually the name of a popular and populist British TV cookery show): *Ready, Steady, Cook!*

BIBLIOGRAPHY

•

Adler, T.A. (1981) 'Making pancakes on Sunday: the male cook in family tradition', *Western Folklore* 40: 45–54.

AgExporter (1989) 'European yuppies – the new niche market for U.S. food exporters?', *AgExporter* December: 4–8.

Ahmad, S., Waller, G. and Verduyn, C. (1994) 'Eating attitudes and body satisfaction among Asian and Caucasian adolescents', *Journal of Adolescence* 17: 461–70.

Alanen, L. (1990) 'Rethinking socialisation, the family and childhood', *Sociological Studies of Child Development* 3: 13–28.

Alexander, R. (1991) 'Men playing the old recipe game', *New York Times* 30 January: C1.

Allen, G. and Crow, G. (1989) *Home and Family*, London: Macmillan.

Anderson, B. (1983) *Imagined Communities: reflections on the origin and spread of nationalism*, London: Verso.

Anderson, D. (ed.) (1986) *A Diet of Reason: sense and nonsense in the healthy eating debate*, London: Social Affairs Unit.

Anon. (1992) 'Confessions of a pregnant woman', *British Medical Journal* 304: 126.

Anthony, A. and Miller, D. (1995) 'What a drag', *Observer* Life 7 May: 26–30.

Appadurai, A. (1981) 'Gastro-politics in Hindu South Asia', *American Ethnologist* 8: 494–511.

—— (1988) 'How to make a national cuisine: cookbooks in contemporary India', *Comparative Studies in Society and History* 30: 3–24.

—— (1990) 'Disjuncture and difference in the global cultural economy', in M. Featherstone (ed.) *Global Culture*, London: Sage.

—— (1993) 'Consumption, duration and history', *Stanford Literature Review* 10: 11–33.

Arce, A. and Marsden, T. (1993) 'The social construction of international food: a new research agenda', *Economic Geography* 69: 293–311.

Ardill, S. (1989) 'Susan Ardill's vegetable curry in coconut milk', in S. O'Sullivan (ed.) *Turning the Tables: recipes and reflections from women*, London: Sheba.

Arreola, D. (1983) 'Mexican restaurants in Tucson', *Journal of Cultural Geography* 3: 108–14.

Atkinson, P. (1983) 'Eating virtue', in A. Murcott (ed.) *The Sociology of Food and Eating*, Aldershot: Gower.

Attar, D. (1985) 'Filthy foreign food', *Camerawork* 31: 13–14.

Ayrton, E. (1980) *English Provincial Cooking*, London: Mitchell Beazley.

Backett, K. (1992) 'Taboos and excesses: lay health moralities in middle class families', *Sociology of Health and Illness* 14: 255–73.

Baker, N. (1986) *The Mezzanine*, London: Penguin.

Bakhtin, M. (1984) *Rabelais and His World*, Bloomington: Indiana University Press.

Ball, R. and Lilly, J. (1982) 'The menace of margarine: the rise and fall of a social problem', *Social Problems* 29: 488–98.

Barthel, D. (1989) 'Modernism and marketing: the chocolate box revisited', *Theory, Culture and Society* 6: 429–38.

Barthes, R. (1972) *Mythologies*, London: Jonathan Cape.

Bateman, M. (1996) 'The first fruits of Kiwi cuisine', *Independent on Sunday Magazine* 30 June: 52–3.

Batty, M. and Barr, B. (1994) 'The electronic frontier: exploring and mapping cyberspace', *Futures* 26: 699–712.

Beardsworth, A. (1990) 'Trans-science and moral panics: understanding food scares', *British Food Journal* 92, 5: 11–17.

Beardsworth, A. and Keil, T. (1990) 'Putting the menu on the agenda', *Sociology* 24: 139–51.

—— (1992) 'The vegetarian option: varieties, conversions, motives and careers', *Sociological Review* 40: 252–93.

Belasco, W. (1989) 'Ethnic fast foods: the corporate melting pot', *Food and Foodways* 2: 1–30.

—— (1993) *Appetite for Change: how the counterculture took on the food industry*, 2nd edition, Ithaca, NY: Cornell University Press.

Bell, D. and Valentine, G. (1995a) 'Food, place and identity', unpublished paper available from the authors.

—— (1995b) 'Queer country: rural lesbian and gay lives', *Journal of Rural Studies* 11: 113–22.

—— (1996) 'Eating identities: food, family, home and the lifecourse', paper presented at the Institute of British Geographers Annual Conference, University of Strathclyde, Glasgow.

Bennett, J. (1943) 'Food and status in a rural society', *American Sociological Review* 8: 561–9.

Benson, R., Thomas, L. and Constantine, E. (1995) 'Slice of life', *The Face* May: 130–5.

Berger, N. (1990) *The School Meals Service*, Plymouth: Northcote House.

Berry, W. (1992) 'The pleasures of eating', in D. Curtin and L. Heldke (eds) *Cooking, Eating, Thinking: transformative philosophies of food*, Bloomington: Indiana University Press.

Bhachu, P. (1995) 'New cultural forms and transnational South Asian women: culture, class, and consumption among British South Asian women in the diaspora', in P. van der Beer (ed.) *Nation and Migration: the politics of space in the South Asian diaspora*, Philadelphia: University of Pennsylvania Press.

Bhadrinath, B. (1990) 'Anorexia nervosa in adolescents of Asian extraction', *British Journal of Psychiatry* 156: 565–8.

Binnie, J. (1995) 'Trading places: consumption, sexuality and the production of queer space', in D. Bell and G. Valentine (eds) *Mapping Desire: geographies of sexualities*, London: Routledge.

—— (1996) 'Coming out of geography: towards a queer epistemology?', *Environment and Planning D: Society and Space* (in press).

Bird, J., Curtis, B., Putnam, T., Robertson, G. and Tickner, L. (eds) (1993) *Mapping the Futures: local cultures, global change*, London: Routledge.

Bishop, P. (1991) 'Constable country: diet, landscape and national identity', *Landscape Research* 16: 31–6.

Blaxter, M. and Paterson, E. (1983) 'The goodness is out of it: the meaning of food to two generations', in A. Murcott (ed.) *The Sociology of Food and Eating*, Aldershot: Gower.

Bloomfield, A.V. (1994) 'Tim Hortons: growth of a Canadian coffee and doughnut chain', *Journal of Cultural Geography* 14: 1–16.

Blythman, J. (1995) 'Fresh fields', *Guardian* Weekend 7 October: 52–3.

Bocock, R. (1993) *Consumption*, London: Routledge.

Bodman, A. (1992) 'Holes in the fabric: more on the master weavers in human geography', *Transactions of the Institute of British Geographers* 17: 108–19.

Bordo, S. (1993) *Unbearable Weight: feminism, western culture and the body*, Berkeley: University of California Press.

Born, M. (1995) 'Auf wiedersehen, Walsall! Brits go abroad at home', *Guardian* 18 November: 31.

Bourdieu, P. (1984) *Distinction: a social critique of the judgement of taste*, London: Routledge.

—— (1991) *Language and Symbolic Power*, Cambridge: Polity Press.

Boym, S. (1994) *Common Places: mythologies of everyday life in Russia*, Cambridge, MA: Harvard University Press.

Brady, J. (1996) 'Delicious doughnuts in Berlin: the dilemma of political community in the age of global capitalism', *Bad Subjects* 25 (electronic journal at: http://english-www.hss.cmu.edu/bs/).

Brannen, J. and Moss, P. (1991) *Managing Mothers: dual earner households after maternity leave*, London: Unwin Hyman.

Brannen, J., Dodd, K., Oakley, A. and Storey, P. (1994) *Young People, Health and Family Life*, Buckingham and Philadelphia: Open University Press.

Broadbent, L. (1995) 'Where dieting is state policy', *Marie Claire* (UK edition) October: 10–16.

Brown, L. and Mussell, K. (eds) (1984) 'Introduction', in L. Brown and K. Mussell (eds) *Ethnic and Regional Foodways in the United States: the performance of group identity*, Knoxville: University of Tennessee Press.

Brownmiller, S. (1984) *Femininity*, London: Hamish Hamilton.

Brumberg, J. (1988) *Fasting Girls: the emergence of anorexia nervosa as a modern disease*, Cambridge, MA: Harvard University Press.

Bryant, C., Courtney, A., Markesberry, B. and DeWalt, K. (1985) *The Cultural Feast: an introduction to food and society*, St Paul: West Publishing Company.

Buie, S. (1996) 'Market as mandala: the erotic space of commerce', *Organization* 3: 225–32.

Bull, N. (1988) 'Studies of the dietary habits, food consumption and nutrient intakes of adolescents and young adults', *World Review of Nutrition and Diet* 57: 24–74.

Burnett, J. (1983) *Plenty and Want: a social history of diet in England from 1815 to the present day*, London: Methuen.

Burton, D. (1993) *The Raj at Table: a culinary history of the British in India*, London: Faber and Faber.

Caglar, A. (1995) 'McDoner: *Doner Kebap* and the social positioning struggle of German Turks', in J. Costa and G. Bamossy (eds) *Marketing in a Multicultural World: ethnicity, nationalism, and cultural identity*, Thousand Oaks, CA: Sage.

Calnan, M. (1990) 'Food and health: a comparison of beliefs and practices in middle class and working class households', in S. Cunningham-Burley and N.P. McKeganey (eds) *Readings in Medical Sociology*, London: Tavistock.

Campbell, C. (1995) 'The sociology of consumption', in D. Miller (ed.) *Acknowledging Consumption: a review of new studies*, London: Routledge.

Cannon, G., Dibb, S., Hanssen, M., Spencer, C. and Ursell, A. (1994) *Food File*, London: Boxtree.

Carey, J. (1995) 'Big Mac versus the little people', *Guardian* 15 April: 23.

Carter, E., Donald, J. and Squires, J. (eds) (1993) *Space and Place: theories of identity and location*, London: Lawrence and Wishart.

Chambers, I. (1994) *Migrancy, Culture, Identity*, London: Routledge.

Chaney, D. (1990) 'Subtopia in Gateshead: the MetroCentre as a cultural form', *Theory, Culture and Society* 7: 49–68.

Chapman, G. and Maclean, H. (1993) ' "Junk food" and "health" food: meanings of food in adolescent women's culture', *Journal of Nutrition Education* 25: 108–13.

Charles, N. and Kerr, M. (1986a) 'Food for feminist thought', *Sociological Review* 34, 1: 537–72.

—— (1986b) 'Eating properly, the family and state benefit', *Sociology* 20: 412–29.

—— (1986c) 'Issues of responsibility and control in the feeding of families', in S. Rodmell and A. Watt (eds) *The Politics of Health Education: raising the issues*, London: Routledge and Kegan Paul.

—— (1988) *Women, Food and Families*, Manchester: Manchester University Press.

Chernin, K. (1992) 'Confessions of an eater', in D. Curtin and L. Heldke (eds) *Cooking, Eating, Thinking: transformative philosophies of food*, Indianapolis: Indiana University Press.

Chetley, A. (1986) *The Politics of Baby Food: successful challenges to an international marketing strategy*, London: Pinter.

Chouinard, V. and Grant, A. (1995) 'On not being even anywhere near "The Project": revolutionary ways of putting ourselves in the picture', *Antipode* 27: 137–66.

Cline, S. (1990) *Just Desserts*, London: André Deutsch.

Cockburn, C. and Fürst-Dilić, R. (eds) (1994) *Bringing Technology Home: gender and technology in a changing Europe*, Buckingham: Open University Press.

Cockburn, C. and Ormrod, S. (1993) *Gender and Technology in the Making*, London: Sage.

Colburn, D. (1992) 'The ideal female body? Thin and getting thinner', *Washington Post* health section, 28 July: 5.

Collins, R. (1985) ' "Horses for courses": ideology and the division of domestic labour', in P. Close and R. Collins (eds) *Family and Economy in Modern Society*, London: Macmillan.

Colomina, B. (1992) 'The split wall: domestic voyeurism', in B. Colomina (ed.) *Sexuality and Space*, Princeton, NJ: Princeton Architectural Press.

Conlin, J.R. (1979) 'Old boy, did you get enough of pie?', *Journal of Forest History* 23: 164–85.

Cook, I. (1994) 'New fruits and vanity: symbolic production in the global food economy', in A. Bonanno, L. Busch, W. Friedland, L. Gouveia and E. Mingione (eds) *From Columbus to ConAgra: the globalization of agriculture and food*, Lawrence: University of Kansas Press.

—— (1995) 'Constructing the exotic: the case of tropical fruit', in J. Allen and C. Hamnett (eds) *A Shrinking World?*, Oxford: Open University Press.

Cook, I. and Crang, P. (1996) 'The world on a plate: culinary culture, displacement and geographical knowledges', *Journal of Material Culture* 1: 131–54.

Corbin, A. (1986) *The Foul and the Fragrant: odour and the French social imagination*, Cambridge, MA: Harvard University Press.

Cosman, M. (1976) *Fabulous Feasts: medieval cookery and ceremony*, New York: George Braziller.

Coupland, D. (1991) *Generation X: tales for an accelerated culture*, New York: St Martin's Press.

Cowan, J. (1991) 'Going out for coffee? Contesting the grounds of gendered pleasures in everyday sociability', in P. Loizos and E. Papataxiarchis (eds) *Contested Identities: gender and kinship in modern Greece*, Princeton, NJ: Princeton University Press.

Coward, R. (1984) *Female Desire: women's sexuality today*, London: Paladin.

Crang, P. (1994) 'It's showtime: on the workplace geographies of display in a restaurant in southeast England', *Environment and Planning D: Society and Space* 12: 675–704.

—— (1995) 'The world on a plate: or, making a meal out of culture, economy, place and space', paper presented during Academic Study Group visit to Israel.

—— (1996) 'Displacement, consumption, and identity', *Environment and Planning A* 28: 47–67.

Cream, J. (1995) 'Re-solving riddles: the sexed body', in D. Bell and G. Valentine (eds) *Mapping Desire: geographies of sexualities*, London: Routledge.

Cresswell, T. (1996) *In Place/Out of Place: geography, ideology and transgression*, Minneapolis: Minnesota Press.

Crisp, A. (1974) 'Primary anorexia nervosa or adolescent weight phobia', *The Practitioner* 212: 17–29.

—— (1980) *Anorexia Nervosa: let me be*, London: Academic Press.

Crotty, P. (1988) 'The disabled in institutions: Transforming functional into domestic modes of food provision', in A. Truswell and M. Wahlqvist (eds) *Food Habits in Australia*, Victoria: Rene Gordon.

Crouch, D. (1989) 'The allotment, landscape and locality: ways of seeing landscape and culture', *Area* 21: 261–7.

—— (1991) 'Allotment culture', *Resurgence* 145: 38–9.

Crouch, D. and Matless, D. (1996) 'Refiguring geography: parish maps of Common Ground', *Transactions of the Institute of British Geographers* 21: 236–55.

Curtin, D. and Heldke, L. (eds) (1992) *Cooking, Eating, Thinking: transformative philosophies of food*, Bloomington: Indiana University Press.

Curtis, B. and Pajaczkowska, C. (1994) '"Getting there": travel, time and narrative', in G. Robertson, M. Mash, L. Tickner, J. Bird, B. Curtis and T. Putnam (eds) *Travellers' Tales: narratives of home and displacement*, London: Routledge.

Daniels, T. and Gerson, J. (eds) (1989) *The Colour Black*, London: British Film Institute.

David, E. (1960) *French Provincial Cooking*, London: Penguin.

Day, H. (1966) *The Complete Book of Curries*, London: Nicholas Kaye.

De Vita, F. (1994) 'The ethnic food business: an overview on the growth of ethnic food market', *Local Economic Quarterly* 3: 90–108.

Dejong, W. (1980) 'The stigma of obesity: the consequences of naive assumptions concerning the causes of physical deviance', *Journal of Health and Social Behaviour* 21: 75–87.

Dejong, W. and Kleck, E. (1986) 'The social psychological effects of overweight', in P. Herman, M. Zanna and E. Higgins (eds) *Physical Appearance, Stigma, and Social Behaviour: the Ontario Symposium*, vol. 3, Hillsdale, NJ: Lawrence Erlbaum.

Delamont, S. (1995) *Appetites and Identities: an introduction to the social anthropology of Western Europe*, London: Routledge.

Delgado, A. (1977) *The Annual Outing and Other Excursions*, London: Allen and Unwin.

Delphy, C. (1979) 'Sharing the same table: consumption and the family', in C. Harris (ed.) *The Sociology of the Family: new directions for Britain*. Sociological Review Monograph 28.

Department of Health (1992) *The Health of the Nation*. London: HMSO.

Department of Health and Social Security (1978) *Eating for Health*, London: HMSO.

DeVault, M.L. (1991) *Feeding the Family: the social organisation of caring and gendered work*, Chicago: University of Chicago Press.

Diamond, R. (1993) 'Become spoiled Moroccan royalty for an evening: the allure of ethnic eateries', *Bad Subjects* 19 (electronic journal at: http://english-www.hss.cmu.edu/bs/).

Dickenson, J. and Salt, J. (1982) 'In vino veritas: an introduction to the geography of wine', *Progress in Human Geography* 6: 159–89.

Dobash, R.E. and Dobash, R. (1980) *Violence against Wives*, London: Open Books.

Dorfman, C. (1992) 'The Garden of Eating: the carnal kitchen in contemporary American culture', *Feminist Issues* 12: 21–38.

Douglas, M. (1970) *Purity and Danger: an analysis of concepts of pollution and taboo*, London: Routledge and Kegan Paul.

—— (1979) 'Les structures du culinaire', *Communications* 31: 145–70.

—— (ed.) (1984) *Food in the Social Order: studies of food and festivities in three American communities*, New York: Russell Sage Foundation.

Douglas, M. and Nicod, M. (1974) 'Taking the biscuit: the structure of British meals', *New Society* 19: 744–7.

Drake, M.A. (1992) 'The nutritional status and dietary inadequacy of single homeless women and their children in shelters', *Public Health Reports* 107: 312–19.

Driver, C. *The British at Table, 1940–1980*, London: Chatto and Windus.

Du Gay, P. (1996) *Consumption and Identity at Work*, London: Sage.

Durham, M. (1996) 'Bring me crocodile and make it snappy', *Observer* Business 19 May: 4.

Dyer, R. (1982) 'Don't look now – the instabilities of the male pin-up', *Screen* 23: 61–73.

Edgell, S. (1980) *Middle Class Couples: a study of segregation, domination and inequality in marriage*, London: Allen and Unwin.

Ehrlich, R. (1995) 'Virtual cookery', *Guardian* Weekend 14 October: 46–7.

Elias, N. (1978) *The Civilising Process*, New York: Urizen (UK edition 1994, Oxford: Blackwell).

Ellis, R. (1982) 'The way to a man's heart: food in the violent home', in A. Murcott (ed.) *The Sociology of Food and Eating*, Aldershot: Gower.

England, K. (1991) 'Gender relations and the spatial structure of the city', *Geoforum* 22: 135–47.

England, P. and Farkas, G. (1986) *Employment, Households and Gender: a social economic and demographic view*, Hawthorne, NY: Aldine de Gruyter.

Enloe, C. (1990) *Bananas, Beaches and Bases: making feminist sense of international politics*, Berkeley: University of California Press.

Esquivel, L. (1993) *Like Water for Chocolate*, London: Black Swan.

Faderman, L. (1991) *Odd Girls and Twilight Lovers: a history of lesbian life in twentieth-century America*, London: Penguin.

Falk, P. (1994) *The Consuming Body*, London: Sage.

Fan, M. (1994) 'Galloping gluttony', *China Now* 149: 24.

Fantasia, R. (1995) 'Fast food in France', *Theory and Society* 24: 201–43.

Featherstone, M. (1987) 'Leisure, symbolic power and the life course', in J. Horne (ed.) *Leisure, Sport and Social Relations*. Sociological Review Monograph 33.

—— (ed.) (1990) *Global Culture*, London: Sage.

—— (1991) *Consumer Culture and Postmodernism*, London: Sage.

Featherstone, M. and Lash, S. (1995) 'Globalization, modernity and the spatialization of social theory: an introduction', in M. Featherstone, S. Lash and R. Robertson (eds) *Global Modernities*, London: Sage.

Featherstone, M., Lash, S. and Robertson, R. (eds) (1995) *Global Modernities*, London: Sage.

Ferguson, P. and Zukin, S. (1995) 'What's cooking', *Theory and Society* 24: 193–9.

Fiddes, N. (1991) *Meat: a natural symbol*, London: Routledge.

Fine, B. (1995) 'From political economy to consumption', in D. Miller (ed.) *Acknowledging Consumption: a review of new studies*, London: Routledge.

Fine, B., Heasman, M. and Wright, J. (1996) *Consumption in the Age of Affluence: the world of food*, London: Routledge.

Fine, B. and Leopold, E. (1993) *The World of Consumption*, London: Routledge.

Fine, G. (1988) 'Letting off steam? Redefining a restaurant's work environment', in M. Jones, M. Moore and R. Snyder (eds) *Inside Organizations: understanding the human dimension*, London: Sage.

—— (1995) 'Wittgenstein's kitchen: sharing meaning in restaurant work', *Theory and Society* 24: 245–69.

Finkelstein, J. (1989) *Dining Out: a sociology of modern manners*, Cambridge: Polity Press.

Fischler, C. (1980) 'Food habits, social change and the nature/culture dilemma', *Social Science Information* 19: 937–53.

—— (1986) 'Learned versus "spontaneous" dietetics: French mothers' views of what children should eat', *Social Science Information* 25: 945–65.

—— (1988) 'Food, self and identity', *Social Science Information* 27: 275–92.

Fisher, M.F.K. (1943) 'The gastronomical me', in *The Art of Eating* (1976), London: Vintage Random.

Fishwick, M. (1983) *Ronald Revisited: the world of Ronald McDonald*, Bowling Green, OH: Popular Press.

Fiske, J. (1989) *Reading the Popular*, London: Routledge.

—— (1993) *Power Plays, Power Works*, London: Verso.

Floyd, K. (1987) *Floyd on France*, London: BBC Books.

Fort, M. (1996a) 'The great grub war – and how we lost it', *Guardian* 10 February: 31.

—— (1996b) 'Well done, that man', *Guardian* Weekend 6 January: 24–5.

Foucault, M. (1977) *Discipline and Punish: the birth of the prison*, London: Tavistock.

Foulkes, N. (1996) 'Eat Me' *Vogue* (UK edition) April: 40–3.

Frank, A. (1991) 'For a sociology of the body: an analytical review', in M. Featherstone, M. Hepworth and B. Turner (eds) *The Body, Social Process and Cultural Theory*, Sage: London.

Frank, J., Fallows, S. and Wheelock, J. (1984) 'The National Food Survey: whose purpose does it serve?', *Food Policy* February: 53–67.

Fraser, W.H. (1981) *The Coming of the Mass Market 1850–1914*, Basingstoke: Macmillan.

Friedman, J. (1990) 'Being in the world: globalization and localization', in M. Featherstone (ed.) *Global Culture*, London: Sage.

—— (1995) 'Global system, globalization and the parameters of modernity', in M. Featherstone, S. Lash and R. Robertson (eds) *Global Modernities*, London: Sage.

Fuss, D. (1989) *Essentially Speaking: feminism, nature, and difference*, New York: Routledge.

Gabriel, Y. and Lang, T. (1995) *The Unmanageable Consumer: contemporary consumption and its fragmentations*, London: Sage.

Galef, D. (1994) 'You aren't what you eat: Anita Brookner's dilemma', *Journal of Popular Culture* 28, 3: 1–7.

Gamman, L. and Makinen, M (1994) *Female Fetishism: a new look*, London: Lawerence and Wishart.

Gannon, M.J. (1994) *Understanding Global Cultures: metaphorical journeys through 17 countries*, Thousand Oaks, CA: Sage.

Garner, D., Garfinkel, P., Schwartz, D. and Thompson, M. (1980) 'Cultural expectations of thinness in women', *Psychological Reports* 47: 483–91.

George, S. (1982) *Food for Beginners*, New York: Writers and Readers.

Gili, E. (1963) *Tia Victoria's Spanish Kitchen*, London: Nicholas Kaye.

Gillespie, A. (1984) 'A wilderness in the megalopolis: foodways in the Pine Barrens of New Jersey', in L.K. Brown and K. Mussell (eds) *Ethnic and Regional Foodways in the United States*, Knoxville: University of Tennessee Press.

Gillespie, M. (1995) *Television, Ethnicity and Cultural Change*, London: Routledge.

Glenn, J. (1995) 'Americana', *Observer* Preview 29 October: 46.

Goffman, E. (1959) *The Presentation of Self in Everyday Life*, New York: Doubleday Books.

—— (1964) *Stigma: notes on the management of spoiled identity*, Englewood Cliffs, NJ: Prentice-Hall.

Gofton, L. (1990) 'Food fears and time famines: some social aspects of choosing and using food', *British Nutrition Foundation Bulletin* 15: 79–95.

Goldman, A. (1992) '"I yam what I yam": cooking, culture, and colonialism', in S. Smith and J. Watson (eds) *De/Colonizing the Subject: the politics of gender in women's autobiography*, Minneapolis: University of Minnesota Press.

Goode, J., Curtis, K. and Theophano, J. (1984) 'Menu formats, meal cycles, and menu negotiation in the maintenance of an Italian-American community', in M. Douglas (ed.) *Food in the Social Order: studies of food and festivities in three American communities*, New York: Russell Sage Foundation.

Goodman, D. and Redclift, M. (1991) *Refashioning Nature: food ecology and culture*, London: Routledge.

Goodman, D., Sorj, B. and Wilkinson, J. (1987) *From Farming to Biotechnology: a theory of agro-industrial development*, Oxford: Blackwell.

Goody, J. (1982) *Cooking, Cuisine and Class*, Cambridge: Cambridge University Press.

Gordon, R.A. (1990) *Anorexia and Bulimia*, Oxford: Blackwell.

Graham, A. (1981) '"Let's eat!" Commitment and communion in co-operative households', *Western Folklore* 40, 1: 55–63.

Graham, H. (1987) 'Being poor: perceptions and coping strategies of lone mothers', in J. Brannen and G. Wilson (eds) *Give and Take in Families*, London: Allen and Unwin.

Green, S. (1991) Making transgressions: the use of style in a women-only community in London, *Cambridge Anthropology* 15: 71–87.

Greenwood, T. and Richardson, D. (1979) 'Nutrition during adolescence', *World Review of Nutrition and Diet*, 33: 1–41.

Gregory, D. (1978) *Ideology, Science and Human Geography*, London: Hutchinson.

Gregson, N. (1995) 'And now it's all consumption?', *Progress in Human Geography* 19: 135–41.

Grigg, D. (1995) 'The pattern of world protein consumption', *Geoforum* 26: 1–17.

Grossberg, L. (1996) 'The space of culture, the power of space', in I. Chambers and L. Curti (eds) *The Post-colonial Question: common skies, divided horizons*, London: Routledge.

Guillen, E. and Barr, S. (1994) 'Nutrition, dieting and fitness messages in a magazine for adolescent women, 1970–1990', *Journal of Adolescent Health* 15: 464–72.

Gutierrez, C.P. (1984) 'The social and symbolic uses of ethnic/regional foodways: Cajuns and crawfish in south Louisiana', in L.K. Brown and K. Mussell (eds) *Ethnic and Regional Foodways in the United States*, Knoxville: University of Tennessee Press.

—— (1992) *Cajun Foodways*, Jackson: University of Mississippi Press.

Haas, R. (1985) *Eat to Win*, Harmondsworth, Middlesex: Penguin.

Haastrup, L. (1992) 'Food cultures, household types and life-modes', *Ethnologia Scandinavica*, 22: 52–66.

Habermas, J. (1962) *The Structural Transformation of the Public Sphere*, Cambridge: Polity Press.

Hall, S. (1991) 'Old and new identities, old and new ethnicities', in A. King (ed.) *Culture, Globalization and the World-System*, Basingstoke: Macmillan.

Hamilton-Paterson, J. (1996) 'The rrruff!! guide to eating', *Guardian* 18 May: 31.

Hannerz, U. (1990) 'Cosmopolitans and locals in world culture', in M. Featherstone (ed.) *Global Culture*, London: Sage.

—— (1996) *Transnation Connections: culture, people, places*, London: Routledge.

Hardyment, C. (1988) *From Mangle to Microwave: the mechanization of household work*, Cambridge: Polity Press.

—— (1995) *Slice of Life: the British way of eating since 1945*, London, BBC Books.

Hartley, J. (1992) *The Politics of Pictures: the creation of the public in the age of popular media*, London: Routledge.

Harvey, D. (1989) *The Condition of Postmodernity*, Oxford: Blackwell.

—— (1990) 'Between space and time: reflections on the geographical imagination', *Annals of the Association of American Geographers* 80: 418–34.

—— (1993) 'Class relations, social justice and the politics of difference', in M. Keith and S. Pile (eds) *Place and the Politics of Identity*, London: Routledge.

Hayden, D. (1980) 'What would a non-sexist city be like? Speculations on housing, urban design and work', *Signs* 5: 171–85.

Heal, C. and Allsop, M. (1986) *Queer Gear: how to buy and cook exotic fruits and vegetables*, London: Century Hutchinson.

Heidenry, C. (1992) *An Introduction to Macrobiotics*, New York: Avery.

Heldke, L. (1992) 'Food politics, political food', in D. Curtin and L. Heldke (eds) *Cooking, Eating, Thinking: transformative philosophies of food*, Bloomington: Indiana University Press.

Helmer, J. (1992) 'Love on a bun: how McDonald's won the burger wars', *Journal of Popular Culture* 26: 85–97.

Henderson, D. (1992) 'Exotic produce: the changing market in the UK', *British Food Journal* 94: 19–24.

Hodgkinson, T. (1995) 'Champagne anarchists', *Observer* Review 10 September: 8.

Hodgson, A. and Bruhn, C. (1993) 'Consumer attitudes toward the use of geographical product descriptors as a marketing technique for locally grown or manufactured foods', *Journal of Food Quality* 16: 163–74.

Hoekveld, G. (1990) 'Regional geography must adjust to new realities', in R. Johnston, J. Hauer and G. Hoekveld (eds) *Regional Geography: current developments and future prospects*, London: Routledge.

Holliday, R. (1995) *Investigating Small Firms: nice work?*, London: Routledge.

—— (1996) 'The workplace turned upside down', paper presented at the American Culture Association/Popular Culture Association conference, Las Vegas, April.

—— (forthcoming) Review of *Consumption and Identity at Work* and *The Unmanageable Consumer*, in *Journal of Management Learning*.

Holt, H. and Pym, H. (eds) (1984) *A Very Private Eye: an autobiography in diaries and letters*, New York: Dutton.

Hopwood, C. (1995) 'My discourse/My-self: therapy as possibility for women who eat compulsively', *Feminist Review* 49: 66–82.

Howes, D. (1990) '"We Are the World" and its counterparts: popular song as constitutional discourse', *Politics, Culture, and Society* 3: 315–39.

Hughes, C. (1991) *Step Parents: wicked or wonderful?*, Aldershot: Avebury Press.

Hughes, G. (1995) 'Authenticity in tourism', *Annals of Tourism Research* 22: 781–803.

Hunt, G. and Satterlee, S. (1986) 'Cohesion and division: drinking in an English village', *Man* 21: 521–37.

Hunter, J. (1993) 'Macroterme geophagy and pregnancy clays in Southern Africa', *Journal of Cultural Geography* 14: 69–92.

Independent (1993a) 'The right vitamin if you can get it', 12 January: 12.

—— (1993b) 'Three cups of coffee a day pose risk to unborn child', 23 December: 6.

Jackson, P. (1994) 'Black male: advertising and the cultural politics of masculinity', *Gender, Place and Culture* 1: 49–59.

Jackson, P. and Holbrook, B. (1996) 'Multiple meanings: shopping and the cultural politics of identity', *Environment and Planning A* 27: 1913–30.

Jackson, P. and Thrift, N. (1995) 'Geographies of consumption', in D. Miller (ed.) *Acknowledging Consumption: a review of new studies*, London: Routledge.

Jackson, S. and Moores, S. (eds) (1995) *The Politics of Domestic Consumption*, London: Prentice-Hall.

Jaivin, L. (1996) *Eat Me*, London: Chatto and Windus.

James, A. (1990) 'The good, the bad and the delicious: the role of confectionery in British society', *Sociological Review* 38: 666–88.

—— (1993) 'Eating green(s): discourses on organic food', in K. Milton (ed.) *Environmentalism: the review from anthropology*, London: Routledge.

James, A. and Prout, A. (eds) (1990) *Constructing and Reconstructing Childhood: contemporary issues in the sociological study of children*, Basingstoke: Falmer Press.

Jansson, S. (1995) 'Food practices and the division of domestic labour: a comparison between British and Swedish households', *Sociological Review* 43: 462–77.

Jary, D. and Jary, J. (eds) (1991) *Collins Dictionary of Sociology*, London: HarperCollins.

Jerome, N.W. (1980) 'Diet and acculturation: the case of Black-American in-migrants', in N. Jerome, R. Kandel and G. Pelto (eds) *Nutritional Anthropology: contemporary approaches to diet and culture*, New York: Redgrave.

—— (1981) 'Frozen (TV) dinners – the staple emergency meals of a changing modern society', in A. Fenton and T.M. Owen (eds) *Food in Perspective*, Edinburgh: John Donald.

Johnson, T. (1979) 'Work together, eat together', in R. Anderson (ed.) *North Atlantic Maritime Cultures*, The Hague: Mouton de Gruyter.

Johnston, L. and Valentine, G. (1995) 'Wherever I lay my girlfriend, that's my home: the performance and surveillance of lesbian identities in domestic environments', in D. Bell and G. Valentine (eds) *Mapping Desire: geographies of sexualities*, London: Routledge.

Johnston, R. (1990) 'The challenge for regional geography: some proposals for research frontiers', in R. Johnston, J. Hauer and G. Hoekveld (eds) *Regional Geography: current developments and future prospects*, London: Routledge.

Johnston, R., Hauer, J. and Hoekveld, G. (1990) 'Region, place and locale: an introduction to different concepts of regional geography', in R. Johnston, J. Hauer and G. Hoekveld (eds) *Regional Geography: current developments and future prospects*, London: Routledge.

Jones, A. (1993) 'Defending the border: men's bodies and vulnerability', *Cultural Studies from Birmingham* 2: 77–123.

Jones, I. (1971) *The Grubbag: an underground cookbook*, cited in W. Belasco (1993) *Appetite for Change: how the counterculture took on the food industry*, 2nd edition, Ithaca, NY: Cornell University Press.

Jones, J. (1996) 'The bad food trap', *Observer* 21 January: 13.

Kalcik, S. (1984) 'Ethnic foodways in America: symbol and the performance of identity', in L.K. Brown and K. Mussell (eds) *Ethnic and Regional Foodways in the United States*, Knoxville: University of Tennessee Press.

Kalka, I. (1991) 'Coffee in Israeli suburbs', *Leisure Studies* 10: 119–31.

Katz, C. and Monk, J. (1993) *Full Circles: geographies of women over the life course*, London: Routledge.

Keane, A. and Willetts, A. (1995) *Concepts of Healthy Eating: an anthropological investigation in south east London*, London: Goldsmiths' College, University of London.

Keith, M. and Pile, S. (eds) (1993) *Place and the Politics of Identity*, London: Routledge.

Keynote (1990) *Confectionary: Keynote Report*, 9th edition, Harlow: Keynote Publications.

—— (1995) *Ethnic Foods: 1995 Market Report*, 6th edition, Harlow: Keynote Publications.

King, A.D. (1990) *World Cities*, London: Routledge.

—— (ed.) (1991) *Culture, Globalization and the World-System*, Basingstoke: Macmillan.

Kirk, M. and Gillespie, A. (1990) 'Factors affecting food choice of working mothers with young families', *Journal of Nutritional Education* 22: 161–8.

Klassen, M., Wauer, S. and Cassel, S. (1990–1) 'Increases in health and weight loss claims in food advertising in the eighties', *Journal of Advertising Research* December/January: 32–7.

Klein, R. (1993) *Cigarettes Are Sublime*, Durham, NC: Duke University Press.

Knabb, K. (ed.) (1989) *Situationist International Anthology*, Berkeley: Bureau of Public Secrets.

Knox, P. (1991) 'The restless urban landscape: economic and sociocultural change and the transformation of metropolitan Washington, DC', *Annals of the Association of American Geographers* 81: 181–209.

—— (1995) 'World cities and the organization of global space', in R. Johnston, P. Taylor and M. Watts (eds) *Geographies of Global Change: remapping the world in the late twentieth century*, Oxford: Blackwell.

Kodras, J. (1992) 'Breadlines', in D. Janelle (ed.) *Geographical Snapshots of North America*, New York: Guilford.

Kramer, J.L. (1995) 'Bachelor farmers and spinsters: gay and lesbian identities and communities in rural North Dakota', in D. Bell and G. Valentine (eds) *Mapping Desire: geographies of sexualities*, London: Routledge.

Kraut, A. (1979) 'Ethnic foodways: the significance of food in the designation of cultural boundaries between immigrant groups in the U.S., 1840–1921', *Journal of American Culture* 2: 409–20.

Lacey, H. (1995) 'Watch out: street party about', *Independent on Sunday* 16 April: 22.

Lanchester, J. (1995a) 'Tales from the salad cart', *Observer* Life 9 July: 45.

—— (1995b) 'The men from down under', *Observer* Life 30 July: 38.

Lane, M. (1995) *Jane Austen and Food*, London: Hambledon Press.

Lang, T. (1986/7) 'The new food policies', *Critical Social Policy* 18: 32–47.

Langdon, P. (1986) *Orange Roofs, Golden Arches: the architecture of American chain restaurants*, London: Michael Joseph.

Lappé, F.M. (1971) *Diet for a Small Planet* (revised edition 1975), New York: Ballatine Books.

Lash, S. and Urry, J. (1994) *Economies of Signs and Space*, London: Sage.

Lawrence, M. (1988) *The Anorexic Experience*, London: Women's Press.

Le Heron, R. and Roche, M. (1995) 'A "fresh" place in food's space', *Area* 27: 22–23.

Leidner, R. (1993) *Fast Food, Fast Talk: the routinization of everyday life*, Berkeley: University of California Press.

Levens, M. (1995) *Eating Disorders and Magical Control of the Body*, London: Routledge.

Levenstein, H. (1988) *Revolution at the Table: the transformation of the American diet*, New York: Oxford University Press.

Lévi-Strauss, C. (1964/1970) *The Raw and The Cooked*, London: Jonathan Cape.

Livingstone, S. (1992) 'The meaning of domestic technologies: a personal construct analysis of familial gender relations', in R. Silverstone and E. Hirsch (eds) *Consuming Technologies: media and information in domestic space*, London: Routledge.

Lloyd, T.C. (1981) 'The Cincinnati chili culinary complex', *Western Folklore* 40: 28–40.

Lo, K. (1994) 'Making a meal of it', *China Now* 149: 16–18.

Longhurst, R. (1995) 'The body in geography', *Gender, Place and Culture* 2: 97–106.

Loveday, L. and Chiba, S. (1985) 'Partaking with the divine and symbolizing the societal: the semiotics of Japanese food and drink', *Semiotics* 56: 115–31.

Lowe, G., Foxcroft, D. and Sibley, D. (1993) *Adolescent Drinking and Family Life*, Reading: Harwood.

Lowe, M. and Wrigley, N. (1996) 'Towards the new retail geography', in N. Wrigley and M. Lowe (eds) *Retailing, Consumption and Capital: towards the new retail geography*, Harlow: Longman.

Lunt, P.K. and Livingstone, S.M. (1992) *Mass Consumption and Personal Identity*, Buckingham: Open University Press.

Lupton, D. (1994) 'Food, memory and meaning: the symbolic and social nature of food events', *Sociological Review* 42, 4: 665–85.

—— (1996) *Food, the Body and the Self*, London: Sage.

Lupton, E. and Abbott Miller, J. (1992) 'Hygiene, cuisine and the product world of early twentieth-century America', in J. Crary and S. Kwinter (eds) *Incorporations*, Cambridge, MA: Zone Books/ MIT Press.

Luxton, M. (1980) *More Than a Labour of Love: three generations of women's work in the home*, Toronto: Women's Educational Press.

MacClancy, J. (1992) *Consuming Culture*, London: Chapman.

McDowell, L. (1983) 'Towards an understanding of the gender division of urban space', *Environment and Planning D: Society and Space* 1: 59–72.

—— (1995) 'Body work: heterosexual performances in city workplaces', in D. Bell and G. Valentine (eds) *Mapping Desire: geographies of sexualities*, London: Routledge.

McIntosh, W.A, Shifflett, P.A. and Picou, J.S. (1989) 'Social support, stressful events, strain, dietary intake and the elderly', *Medical Care* 27: 140–53.

MacIntyre, S. (1983) 'The management of food in pregnancy', in A. Murcott (ed.) *The Sociology of Food and Eating*, Aldershot: Gower.

McKie, L., Wood, R. and Gregory, S. (1993) 'Women defining health: food, diet and body image', *Health Education Research* 8: 35–41.

McMichael, P. and Myrhe, D. (1991) 'Global regulation vs. the nation-state: agro-food systems and the new politics of capital', *Capital and Class* 43: 83–105.

McSween, M. (1993) *Anorexic Bodies*, London: Routledge.

Mansfield, P. and Collard, J. (1988) *The Beginning of the Rest of Your Life? A portrait of newly-wed marriage*, Basingstoke: Macmillan.

Mars, G. (1987) 'Longshore drinking, economic security and union politics in Newfoundland', in M. Douglas (ed.) *Constructive Drinking: perspectives on drink from anthropology*, Cambridge: Cambridge University Press.

Mars, G. and Nicod, M. (1984) *The World of Waiters*, London: Allen and Unwin.

Martin, E. (1992) 'The end of the body?', *American Ethnologist* 19: 121–40.

Massey, D. (1993) 'Power-geometry and a progressive sense of place', in J. Bird, B. Curtis, T. Putnam, G. Robertson and L. Tickner (eds) *Mapping the Futures: local cultures, global change*, London: Routledge.

—— (1995) 'Places and their pasts', *History Workshop Journal* 39: 182–92.

Mathews, V. and Westie, C. (1966) 'A preferred method for obtaining rankings: reactions to physical handicaps', *American Sociological Review* 31: 47–63.

Matrix (1984) *Making Space: women and the man made environment*, London: Pluto Press.

May, J. (1993) 'Constructions of the "exotic" and the "authentic" in the everyday', paper presented at the Institute of British Geographers annual conference, Royal Holloway and Bedford New College, London, January.

Maynard, M. (1995) 'Unclaimed bananas: the signification of devalued food in a public space', paper presented at the American Culture Association/Popular Culture Association conference, Philadelphia, March.

Mead, M. (1980) 'A perspective on food patterns', in L. Tobias and P. Thompson (eds) *Issues in Nutrition from the 1980s*, Monterey: Wadsworth.

Mechling, J. and Wilson, D. (1988) 'Organizational festivals and the uses of ambiguity: the case of Picnic Day at Davis', in M. Jones, M. Moore and R. Snyder (eds) *Inside Organizations: understanding the human dimension*, London: Sage.

Mennell, S. (1985) *All Manners of Food: eating and taste in England and France from the Middle Ages to the present*, Oxford: Blackwell.

—— (1991) 'On the civilising of appetite', in M. Featherstone, M. Hepworth and B. Turner (eds) *The Body: social process and cultural theory*, London: Sage.

Mennell, S., Murcott, A. and van Otterloo, A. (1992) *The Sociology of Food: eating, diet and culture*, London: Sage.

Middleton, S. and Thomas, M. (1994) 'Saying "No" or giving in gracefully', in S. Middleton, K. Ashworth and R. Walker (eds) *Family Fortunes*, London: Child Poverty Action Group.

Miles, E. (1993) 'Adventures in the postmodernist kitchen: the cuisine of Wolfgang Puck', *Journal of Popular Culture* 27: 191–203.

Miller, D. (1995) 'Consumption as the vanguard of history: a polemic by way of an introduction', in D. Miller (ed.) *Acknowledging Consumption: a review of new studies*, London: Routledge.

Miller, R. (1983) 'The Hoover® in the garden: middle-class women and suburbanization, 1850–1920', *Environment and Planning D: Society and Space* 6: 191–212.

Mintz, S. (1984) 'Meals without grace', *Boston Review* December: 6–7.

—— (1985) *Sweetness and Power: the place of sugar in modern history*, New York: Viking.

Mitchell, V.W. and Greatorex, M. (1990) 'Consumer perceived risk in the UK food market', *British Food Journal* 92: 16–22.

Moir, J. (1995) 'Dispatches from the front line of supermarket shopping', *Observer* Life 30 July: 5.

—— (1996) 'But there *is* a fly in my soup', *Observer* Review 16 June: 7.

Montanari, M. (1994) *The Culture of Food*, Oxford: Blackwell.

Moran, W. (1993) 'Rural space as intellectual property', *Political Geography* 12: 263–77.

Morgan, D.H.J. (1996) *Family Connections*, Cambridge: Polity Press.

Morley, D. (1991) 'Where the global meets the local: notes from the sitting room', *Screen* 32: 1–15.

Morris, A., Cooper, T. and Cooper, P. (1989) 'The changing face of female fashion models', *International Journal of Eating Disorders* 8: 593–6.

Morrison, M. (1995) 'Researching food consumers in schools: recipes for concern', *Educational Studies* 21: 239–63.

Morse, M. (1994) 'What do cyborgs eat? Oral logic in an information society', in G. Bender and T. Druckrey (eds) *Culture on the Brink: ideologies of technology*, Boston: Bay Press.

Mort, F. (1988) 'Boy's own? Masculinity, style and popular culture', in R. Chapman and J. Rutherford (eds) *Male Order: unwrapping masculinity*, London: Lawrence & Wishart.

—— (1996) *Cultures of Consumption: masculinities and social space in late twentieth-century Britain*, London: Routledge.

Mumford, D., Whitehouse, A. and Platts, M. (1991) 'Socio-cultural corrolates of eating disorders among Asian schoolgirls in Bradford', *British Journal of Psychiatry* 158: 173–84.

Munro, R. (1995) 'Extension, exchange and identity: the consumption view of the self', paper presented at ESRC Consumption Seminar Series. Available from the author: Centre for Social Theory and Technology, Keele University, Staffordshire.

Murcott, A. (1982a) 'Menus, meals and platefuls: observations on advice about diet in pregnancy', *International Journal of Sociology* 2: 1–11.

—— (1982b) 'On the social significance of the "cooked dinner" in South Wales', *Social Science Information* 21: 677–95.

—— (1983a) 'Cooking and the cooked: a note on the domestic preparation of meals', in A. Murcott (ed.) *The Sociology of Food and Eating*, Aldershot: Gower.

—— (1983b) ' "It's a pleasure to cook for him": food, mealtimes and gender in some South Wales households', in E. Garmarnikow, D. Morgan, J. Purvis and D. Taylorson (eds) *The Public and the Private*, London: Heinemann.

—— (1986) 'Opening the "black box": food, eating and household relationships', *Sosiaalilaaketie-teellinen Aikakuslehti* 23: 85–92.

—— (1993a) 'Purity and pollution: body management and the social place of infancy', in S. Scott and D. Morgan (eds) *Body Matters*, London: Falmer Press.

—— (1993b) 'Talking of good food: an empirical study of women's conceptualisations', *Food and Foodways* 5, 3: 305–18.

Narayan, U. (1995) 'Eating cultures: incorporation, identity and Indian food', *Social Identities* 1: 63–86.

Naughton, J. (1996) 'Veggies get their pound of flesh', *Observer* Review 5 May: 13.

Nederveen Pieterse, J. (1995) 'Globalization as hybridization', in M. Featherstone, S. Lash and R. Robertson (eds) *Global Modernities*, London: Sage.

Newman, L. (1993) *Eating Our Hearts Out: personal accounts of women's relationship to food*, Freedom, CA: The Crossing Press.

Nicod, M. (1974) 'A method of eliciting the social meanings of food', unpublished MA thesis, University College London.

Oakely, A. (1974) *The Sociology of Housework*, Oxford: Martin Robertson.

Ohnuki-Tierney, E. (1993) *Rice as Self: Japanese identities through time*, Princeton, NJ: Princeton University Press.

Orbach, M. (1977) *Hunters, Seamen and Entrepreneurs: the tuna fishermen of San Diego*, Berkeley: University of California Press.

Orbach, S. (1988) *Fat Is a Feminist Issue*, London: Arrow Books.

O'Sullivan, S. (ed.) (1989) *Turning the Tables: recipes and reflections from women*, London: Sheba.

Oulton, C. and Narayan, N. (1995) 'RSPCA accused over Freedom Food', *Observer* 25 June: 5.

Oxford, E. (1995) 'Pizza: a slice of lifestyle', *Independent* 15 May: 2.

Parker, M. and Jary, D. (1995) 'The McUniversity: organization, management and academic subjectivity', *Organization* 2: 319–38.

Paulson-Box, E. and Williamson, P. (1990) 'The development of the ethnic food market in the UK', *British Food Journal* 92: 10–15.

Penn, C. and Wallace, W. (1995) 'The garden of England', *Reportage* 8: 20–5.

Perry, N. (1995) 'Travelling theory/nomadic theorizing', *Organization* 2: 35–54.

Philo, C. (1992) 'Neglected rural geographies: a review', *Journal of Rural Studies* 8: 193–207.

Pillsbury, R. (1990) *From Boarding House to Bistro: the American restaurant then and now*, London: Unwin Hyman.

The Player (pseud.) (1995) 'Babylon', *Observer* Review 22 October: 7.

Popham, P. (1995) 'The great escape', *Independent* 22 September 2–3.

Porter, H. (1996) 'From lunch to bed in three easy courses', *Telegraph Magazine* 6 January: 32–4.

Posener, J. (1987) 'Snow pea fried chicken', in S. O'Sullivan (ed.) *Turning the Tables: reflections and recipes from women*, London: Sheba.

Prattala, R. (1989) *Young People and Food: socio-cultural studies of food consumption patterns*, Helsinki: University of Helsinki.

Prout, A. (1991) Review of *Women, Food and Families*, *Sociological Review* 39: 403–5.

Putnam, T. and Newton, C. (1990) *Household Choices*, London: Futures Publications.

Raspa, R. (1984) 'Exotic foods among Italian-Americans in Mormon Utah: food as nostalgic enactment of identity', in L.K. Brown and K. Mussell (eds) *Ethnic and Regional Foodways in the United States*, Knoxville: University of Tennessee Press.

Redhead, S. (1995) *Unpopular Cultures: the birth of law and popular culture*, Manchester: Manchester University Press.

Rezek, P. and Leary, M. (1991) 'Perceived control, drive for thinness and food consumption: anorexic tendencies as displaced reactance', *Journal of Personality* 59: 129–42.

Reynolds, M. (ed.) (1990) *Erotica: an anthology of women's writing*, London: Pandora.

Rich, A. (1986) *Blood, Bread and Poetry*, London: W.W. Norton & Co.

Ritzer, G. (1993) *The McDonaldization of Society*, Newbury Park, CA: Pine Forge.

Roberts, M. (1991) *Living in a man-made world: gender assumptions in modern housing design*, London: Routledge.

Robertson, G., Mash, M., Tickner, L., Bird, J. and Putnam, T. (1994a) 'As the world turns: introduction', in G. Robertson, M. Mash, L. Tickner, J. Bird, B. Curtis and T. Putnam (eds) *Travellers' Tales: narratives of home and displacement*, London: Routledge.

Robertson, R. (1995) 'Glocalization: time–space and homogeneity–heterogeneity', in M. Featherstone, S. Lash and R. Robertson (eds) *Global Modernities*, London: Sage.

Rogerson, C.M. (1986) 'A strange case of beer: the state and sorghum beer manufacture in South Africa', *Area* 18: 15–24.

Rooke, M. (1995) 'Food fights', *Options* March: 30–33.

Rose, G. (1993) *Feminism and Geography: the limits to geographical knowledge*, Cambridge: Polity Press.

Rose, R. and Falconer, P. (1992) 'Individual taste or effective decision? Public policy on school meals', *Journal of Social Policy* 21: 349–73.

Rosen, M. (1985) 'Breakfast at Spiro's: dramaturgy and dominance', *Journal of Management* 11: 31–48.

——— (1988) 'You asked for it: Christmas at the bosses' expense', *Journal of Management Studies* 25: 463–80.

Ross, K. (1995) *Fast Cars, Clean Bodies; decolonization and the reordering of French Culture*, Cambridge, MA: MIT Press.

Rozin, P. (1987) 'Sweetness, sensuality, sin, safety and socialization: some speculations', in J. Dobbing (ed.) *Sweetness*, London: Springer-Verlag.

Rubenstein, H. (1992) 'The hunger', *Interview* December: 145–6.

Runciman, D. (1996) 'Counter culture', *Guardian* 24 April: 2–3.

Sack, R. (1992) *Place, Modernity, and the Consumer's World*, Baltimore: Johns Hopkins University Press.

Sainsbury's Magazine (1995) 'Food – a consuming passion', August: 54–8.

Samuel, R. (1989) 'Introduction', in R. Samuel (ed.) *Patriotism*, London: Routledge.

Sanjur, D. (1982) *Social and Cultural Perspectives in Nutrition*, Englewood Cliffs, NJ: Prentice-Hall.

Saunders, P. (1989) 'The meaning of "home" in contemporary English culture', *Housing Studies* 4: 177–92.

Schivelbusch, W. (1992) *Tastes of Paradise: a social history of spices, stimulants, and intoxicants*, New York: Pantheon.

Schwartz, H. (1986) *Never Satisfied: a cultural history of diets, fantasies and fat*, London: Collier Macmillan.

Scola, R. (1975) 'Food markets and shops in Manchester 1770–1870', *Journal of Historical Geography* 1: 153–68.

Self, W. (1995a) 'National Griddle', *Observer* Life 24 December: 32.

——— (1995b) 'Reach for the Thai', *Observer* Life 12 November: 57.

Sennett, R. (1994) *Flesh and Stone: the body and the city in western civilization*, London: Faber and Faber.

Shafer, E., Shafer, R., Bulten, G. and Hoiberg, E. (1993) 'Safety of the U.S. food supply: consumer concerns and behaviour', *Journal of Consumer Studies and Home Economics* 17: 137–44.

Sharman, A. (1991) 'From generation to generation: resources, experience, and orientation in the dietary patterns of selected urban American households', in A. Sharman, J. Theophano, K. Curtis and E. Messer (eds) *Diet and Domestic Life in Society*, Philadelphia: Temple University Press.

Shaw, G. (1985) 'Changes in consumer demand and food supply in nineteenth-century British cities', *Journal of Historical Geography* 11: 280–96.

Shapiro, L. (1986) *Perfection Salad: women and cooking at the turn of the century*, New York: Farrar, Straus and Giroux.

Shields, R. (1989) 'Social spatialization and the built environment: West Edmonton Mall', *Environment and Planning D: Society and Space* 7: 147–64.

—— (1991) *Places on the Margin*, London: Routledge.

Shilling, C. (1993) *The Body and Social Theory*, London: Sage.

Shurmer-Smith, P. and Hannam, K. (1994) *Worlds of Desire, Realms of Power: a cultural geography*, London: Edward Arnold.

Shute, J. (1993) *Life Size*, London: Mandarin Paperbacks.

Sibley, D. (1991) 'The fear of others: the boundaries of self, difference and social space', paper presented at the Soviet–British seminar, available from the author at the Department of Geography and Earth Resources, University of Hull.

Silverstein, B., Perdue, L., Peterson, B. *et al.* (1986) 'The role of the mass media in promoting a thin standard of bodily attractiveness for women', *Sex Roles* 14: 519–32.

Silverstone, R. and Hirsch, E. (eds) (1992) *Consuming Technologies: media and information in domestic spaces*, London: Routledge.

Simmonds, D. (1990) 'What's next? Fashion, foodies and the illusion of freedom', in A. Tomlinson (ed.) *Consumption, Identity and Style: marketing, meanings and the packaging of pleasure*, London: Routledge.

Singer, E.A. (1984) 'Conversion through foodways enculturation: the meaning of eating in an American Hindu sect', in L.K. Brown and K. Mussell (eds) *Ethnic and Regional Foodways in the United States*, Knoxville: University of Tennessee Press.

Slater, J.M. (ed.) (1991) *Fifty Years of the National Food Survey, 1940–1990*, London: HMSO.

Slater, N. (1995a) '1996 predictions: food', *Observer* Life 31 December: 19.

—— (1995b) 'Strange fruit', *Observer* Life 9 July: 42–3.

Smart, B. (1994) 'Digesting the modern diet: gastro-porn, fast food and panic eating', in K. Tester (ed.) *The Flâneur*, London: Routledge.

Smith, N. (1993) 'Homeless/global: scaling places', in J. Bird, B. Curtis, T. Putnam, G. Robertson and L. Tickner (eds) *Mapping the Futures: local cultures, global change*, London: Routledge.

Smith, N.C. (1990) *Morality and the Market: consumer pressure for corporate accountability*, London: Routledge.

Smith, W.D. (1992) 'Complications of the commonplace: tea, sugar, and imperialism', *Journal of Interdisciplinary History* 23: 259–78.

Sommerville, P. (1992) 'Homelessness and the meaning of home: rooflessness or rootlessness?', *International Journal of Urban and Regional Research* 16: 528–39.

Spencer, C. (1994) *The Heretic's Feast: a history of vegetarianism*, London: Fourth Estate.

Stacey, J. (1990) *Brave New Families: stories of domestic upheaval in late twentieth century America*, New York: Basic Books.

Star, S.L. (1991) 'Power, technology and the phenomenology of conventions: on being allergic to onions', in J. Law (ed.) *A Sociology of Monsters: essays on power, technology and domination*, London: Routledge.

Stead, J. (1991) 'Prodigal frugality: Yorkshire pudding and parkin, two traditional Yorkshire foods', in C.A. Wilson (ed.) *Traditional Food: east and west of the Pennines*, Edinburgh: Edinburgh University Press.

Steiner, R. (1986) 'Drinking-place names in the central United States', *Journal of Cultural Geography* 6: 19–34.

Stephens, D.L., Hill, R.P. and Hanson, C. (1994) 'The beauty myth and female consumers: the controversial role of advertising', *Journal of Consumer Affairs* 28: 137–53.

Stephenson, D. (1996) 'Food media: globalization, consumerism and identity', unpublished paper, Department of Cultural Studies, Staffordshire University.

Strasser, S. (1982) *Never Done: a history of American housework*, New York: Pantheon Books.

Sullivan, C. (1996) 'Eat, drink, smoke cigars and make whoopee', *Guardian* 15 June: 27.

Sutherland, C., Williams, J. and Mather, C. (1986) 'Beer houses: an indicator of cultural change in Taiwan', *Journal of Cultural Geography* 6: 35–50.

Swinney, B. (1993) *Eating Expectantly: the essential eating guide and cookbook for pregnancy*, Colorado Springs, CO: Fall River Press.

Symons, M. (1983) 'Australia's "one continuous picnic"', in *Food in Motion: the migration of foodstuffs and cookery techniques*, proceedings of the Oxford Symposium, vol. 1, London: Prospect Books.

Synott, A (1993) *The Body Social: symbolism, self and society*, London: Routledge.

Tansey, G. and Worsely, T. (1995) *The Food System: a guide*, London: Earthscan.

Tarantino, Q. (1994) *Pulp Fiction: three stories . . . about one story . . .*, London: Faber and Faber.

Tarrant, J. (1985) 'A review of the international food trade', *Progress in Human Geography* 9: 235–54.

Taylor, I., Evans, K. and Fraser, P. (1996) *A Tale of Two Cities: a study in Manchester and Sheffield*, London: Routledge.

Telfer, E. (1996) *Food for Thought: philosophy and food*, London: Routledge.

Thomas, A. (1978) 'Class and sociability among urban workers: a study of the bar as social club', *Medical Anthropology* 2: 9–30.

Thompson, E.P. (1967) 'Time, work-discipline and industrial capitalism', *Past and Present* 38: 56–97.

Thrall, C. (1982) 'The conservative use of modern household technology', *Technology and Culture* 23: 175–94.

Thrift, N. (1990a) 'Doing regional geography in a global system: the new international finance system, the City of London, and the south east of England, 1984–7', in R. Johnston, J. Hauer and G. Hoekveld (eds) *Regional Geography: current developments and future prospects*, London: Routledge.

—— (1990b) 'For a new regional geography 1', *Progress in Human Geography* 14: 272–9.

—— (1991) 'For a new regional geography 2', *Progress in Human Geography* 15: 456–65.

—— (1993) 'For a new regional geography 3', *Progress in Human Geography* 17: 92–100.

Tomlinson, A. (ed.) (1990) *Consumption, Identity, and Style: marketing, meanings, and the packaging of pleasure*, London: Routledge.

Townley, B. (1994) *Reframing Human Resource Management: power, ethics and the subject at work*, London: Sage.

Tredre, R. (1995) 'Pukka masters of balti cast their chilly gaze south', *Observer* 7 May: 11.

—— (1996) 'Fillet of emu and a pound of 'roo sausages, please', *Observer* Life 10 March : 26–8.

True Light Beaver (1972) *Eat, Fast, Feast*, cited in W. Belasco, *Appetite for Change: how the counterculture took on the food industry*, 2nd edition, Ithaca, NY: Cornell University Press.

Tuan, Y.-F. (1974) *Topophilia: a study of environmental perception, attitudes and values*, Englewood Cliffs, NJ: Prentice-Hall.

Turner, B. (1984) *The Body and Society: explorations in social theory*, Oxford: Basil Blackwell.

—— (1991) 'Recent development in the theory of the body', in M. Featherstone, M. Hepworth and B. Turner (eds) *The Body: social process and cultural theory*, London: Sage.

—— (1992) *Regulating Bodies: essays on medical sociology*, London: Routledge.

Twigg, J. (1983) 'Vegetarianism and the meaning of meat', in A. Murcott (ed.) *The Sociology of Food and Eating*, Aldershot: Gower.

Ufkes, F.M. (1993) 'The globalization of agriculture', *Political Geography* 12: 194–7.

Unwin, T. (1991) *Wine and the Vine: an historical geography of viticulture and the wine trade*, London: Routledge.

Urry, J. (1990) *The Tourist Gaze: leisure and travel in contemporary societies*, London: Sage.

van den Berghe, P. (1984) 'Ethnic cuisine, culture in nature', *Ethnic and Racial Studies* 7: 387–97.

van Otterloo, A. (1987) 'Foreign immigrants and the Dutch at table, 1945–1985. Bridging or widening the gap?', *Netherlands Journal of Sociology* 23: 126–43.

Vardy, W. (1996) 'The briar around the strawberry patch: toys, women and food', *Women's Studies International Forum* 19: 267–76.

Veness, A. (1994) 'Designer shelters as models and makers of home: new responses to homelessness in urban America', *Urban Geography* 15: 150–67.

Visser, M. (1986) *Much Depends on Dinner: the extraordinary history and mythology, allure and obsessions, perils and taboos, of an ordinary meal*, Toronto: McClelland and Stewart.

—— (1991) *The Rituals of Dinner: the origins, evolution, eccentricities and meaning of table manners*, London: Grove Widenfeld.

Waksler, F. (1986) 'Studying children: phenomenological insights', *Human Studies* 8: 171–82.

Walker, R., Ashworth, K., Kellard, K., Middletone, S., Peaker, A. and Thomas, M. (1994) 'Pretty, pretty, please – just like a parrot: persuasion strategies used by children and young people', in S. Middletone, K. Ashworth and R. Walker (eds) *Family Fortunes*, London: Child Poverty Action Group.

Warde, A. (1994) 'Consumption, identity-formation and uncertainty', *Sociology* 28: 877–98.

Warde, A. and Hetherington, K. (1994) 'English households and routine food practices: a research note', *Sociological Review* 42: 758–78.

Waters, M. (1995) *Globalization*, London: Routledge.

Watson, B. (1962) *Cooks, Gluttons and Gourmets*, New York: Garden City.

Weaver, W. (1986) 'White gravies in the American popular diet', in A. Fenton and E. Kisban (eds) *Food in Change: eating habits from the Middle Ages to the present day*, London: John Donald Publishers.

Weightman, B. (1980) 'Gay bars as private places', *Landscape* 23: 9–16.

Westwood, S. (1984) *All Day Every Day: factory and family in the making of women's lives*, London: Pluto Press.

Weyland, P. (1993) *Inside the Third World Village*, London: Routledge.

White, P. and Gillett, J. (1994) 'Reading the muscular body: a critical decoding of advertisements in *Flex* magazine', *Sociology of Sport Journal*, 11: 18–39.

Whitehead, T. (1984) 'Socio-cultural dynamics and food habits in a Southern community', in M. Douglas (ed.) *Food in the Social Order: studies of food and festivities in three American communities*, New York: Russell Sage Foundation.

Widdowson, J. (1981) 'Food and traditional verbal modes in the social control of children', in A. Fenton and T. Owen (eds) *Food in Perspective*, Edinburgh: John Donald.

Wiecha, J., Dwyer J., Jacques, P. and Rand, W. (1993) 'Nutritional and economic advantages for homeless families in shelters providing kitchen facilities and food', *Journal of the American Dietetic Association* 93: 777–83.

Willetts, A. and Keane, A. (1995) 'You eat what you are . . .', *Guardian* Weekend 15 April: 42–3.

Williams, B. (1984) 'Why migrant women feed their husbands tamales: foodways as a basis for a revisionist view of Tejano family life', in L. Brown and K. Mussell (eds) *Ethnic and Regional Foodways in the United States: the performance of group identity*, Knoxville: University of Tennessee Press.

Williamson, B. (1982) *Class, Culture and Community: a biographical study of social change in mining*, London: Routledge and Kegan Paul.

Willis, R. (ed.) (1990) *Signifying Animals: human meaning in the natural world*, London: Routledge.

Willis, S. (1991) *A Primer for Daily Life*, London: Routledge.

Wilson, E. (1988) *Hallucinations: life in the postmodern city*, London: Hutchinson Radius.

—— (1996) 'The cafe: the ultimate bohemian space', in I. Borde, J. Kerr, A. Pivaro and J. Rendell (eds) *Strangely Familiar: narratives of architecture in the city*, London: Routledge.

Wirth, L. (1938) 'Urbanism as a way of life', *American Journal of Sociology* 44: 121–85.

Wiseman, C., Gray, J. and Mosimann, J. (1992) 'Cultural expectations of thinness in women: an update', *International Journal of Eating Disorders* 11: 85–9.

Wolf, N. (1991) *The Beauty Myth: how images of beauty are used against women*, New York: Anchor Books.

Wood, R. (1992) 'Dining out in the urban context', *British Food Journal* 94: 3–5.

—— (1995) *The Sociology of the Meal*, Edinburgh: Edinburgh University Press.

Woodhead, D. (1995) '"Surveillant gays": HIV, space and the constitution of identities', in D. Bell and G. Valentine (eds) *Mapping Desire: geographies of sexualities*, London: Routledge.

Worsley, P. (1987) *New Introductory Sociology*, 3rd edition, London: Penguin.

Wrigley, N. (1991) 'Is the "golden age" of British grocery retailing at a watershed?', *Environment and Planning A* 23: 1537–44.

—— (1987) 'The concentration of capital in UK grocery retailing', *Environment and Planning* A 19: 1283–8.

—— (1989) 'The lure of the USA: further reflections on the internationalisation of British grocery retailing capital', *Environment and Planning A* 21: 283–8.

—— (1996) 'Sunk costs and corporate restructuring: British food retailing and the property crisis', in N. Wrigley and M. Lowe (eds) *Retailing, Consumption and Capital: towards the new retail geography*, Harlow: Longman.

Wrigley, N. and Lowe, M. (eds) (1996) *Retailing, Consumption and Capital: towards the new retail geography*, Harlow: Longman.

Wroe, M. (1995) 'My lords, ladies and gentlemen of the road', *Observer* 5 June: 2.

Young, J. (1994) 'Kitchen life', *Elle Decoration* 'Living Kitchen' supplement: 4–5.

Young, L. (1994) 'Paupers, property, and place: a geography of the English, Irish, and Scottish Poor Laws in the mid-nineteenth century', *Environment and Planning D: Society and Space* 12: 325–40.

—— (forthcoming) *The Geography of Hunger*, London: Routledge.

Young, M. and Wilmott, P. (1975) *The Symmetrical Family*, Harmondsworth, Middlesex: Penguin.

Zelinsky, W. (1985) 'The roving palate: North America's ethnic cuisines', *Geoforum* 16: 51–72.

Zukin, S. (1991) *Landscapes of Power: from Detroit to Disney World*, Berkeley: University of California Press.

—— (1995) *The Cultures of Cities*, Oxford: Blackwell.

INDEX

•